Handbook
for
Sound Technicians

Handbook
for
Sound Technicians

F.P. Zantis

Elektor Electronics (Publishing)

Elektor Electronics (Publishing)
P.O. Box 1414 •Dorchester • England DT2 8YH

British Library Cataloguing in Publication Data
A catalogue record for this book is available from the British Library

ISBN 0 905705 48 3

Translation and make-up: A.W. Moore, MSc

First published in the United Kingdom 1999

© Segment BV 1999

Printed in the Netherlands by Giethoorn/Ten Brink, Meppel

Table of contents

Other books from Elektor Electronics

301 Circuits (ISBN 0 905705 12 2)

302 Circuits (ISBN 0 905705 25 4)

303 Circuits (ISBN 0 905705 26 2)

304 Circuits (ISBN 0 905705 34 3)

305 Circuits (ISBN 0 905705 36 X)

306 Circuits (ISBN 0 905705 43 2)

Build your own AF valve amplifiers (ISBN 0 905705 39 4)

Build your own Electronic Test Instruments (ISBN 0 905705 37 8)

Build your own high-end audio equipment (ISBN 0 905705 40 8)

Designing Audio Circuits (ISBN 0 905705 50 5)

Faultfinding in Computers & Digital Circuits (0 905705 60 2)

I^2C Bus (ISBN 0 905705 47 5)

Lasers: Theory and Practice (ISBN 0 905705 52 1)

Microprocessor Data Book (ISBN 0 905705 28 9)

Matchbox Single-board computer (ISBN 0 905705 53 X)

PC Service & Repair (ISBN 0 905705 41 6)

PICs in Practice (ISBN 0 905705 51 3)

SCSI: The Ins and Outs (ISBN 0 905705 44 0)

SMT Projects (ISBN 0 905705 35 1)

ST62 Microcontrollers (ISBN 0 905705 42 4)

The CD-ROM System (ISBN 0 905705 46 7)

1. Basic theory

A sound engineer cannot work effectively without a basic knowledge of electricity, acoustics and music, particularly so since electrical and acoustical technology do not stand still. This is, of course, also true for a musician who during rehearsals depends on his/her own knowledge. Only someone who is fully acquainted with electro-acoustical technology is able to use the available sound equipment effectively and efficiently, and who can trace and solve any problems quickly and satisfactorily. This is the reason that in this book more attention is paid to the principles of electro-acoustics than to technical details, which may be superseded in the very near future. It is only when the principles are fully comprehended that we are able to understand, judge and apply new techniques, procedures and equipment. Detailed mathematical treatment has been avoided as much as possible.

1.1 Principles of electricity

Sound engineering as we know it would not be possible without electrical energy. Owing to the rapid developments in the fields of electrical and electronics engineering in the past fifty years or so, a number of distinct disciplines has arisen. One of these is electro-acoustics, which deals with the conversion, processing and reproducing of sound, that is, fluctuations of air pressure. This implies that a sound engineer must have at least a basic knowledge not only of acoustics (and music), but also of electrical engineering and electronics. He/she must be familiar with direct and alternating voltages and currents, low-frequency techniques and measurement techniques to be able to thoroughly understand and evaluate the operation of electro-acoustical equipment. This is true not only of the sound engineer, but also of the performing artist. A sound engineer accompanying a travelling group, band or orchestra must be able to trace and repair simple electrical defects in amplifiers, mixers, effects units, lighting equipment, and so on.

1.1.1 Voltage and voltage generation

In an electrical conductor such as a copper wire, freely moving charge carriers, that is, electrons, are distributed evenly. This even distruibution may be upset by external influences. A battery, for instance, by means of a chemical reaction produces a surplus of electrons at its positive terminal and an equally large shortage of electronics at its

1

negative terminal. Since opposite charges attract one another, there is a tendency for the two charges to be partially interchanged. This tendency is called electrical voltage or potential. The level of this voltage depends on the difference in charges at the two terminals: it is high when the difference is large and low when the difference is small. The symbol for voltage or potential is U. The unit of electrical potential or voltage is the volt, V, after the Italian physicist Alessandro Volta. An electrical voltage may be generated in a number of ways , some of wh ich are discussed below.

Voltage generation by means of inductance
Inductance is an electrical property that exists owing to the presence of a magnetic field around a conductor through which a current flows. It is a kind of electrical momentum. A single, straight wire has inductance, but this is very small for most purposes. The inductance may be increased by winding a coil around a magnetic material. The increase in inductance is directly proportional to the square of the number of turns. The inductance, L, of a long coil is

$$L = \mu n^2 A/l, \qquad\qquad\qquad \text{[Eq. 1a]}$$

where L is the inductance in henrys, A is the cross-sectional area of the coil in square metres, l is the length of the coil in metres and μ is the magnetic permeability of the coil's core (= $4\pi \times 10^{-7}$ for air).

When such a coil is placed in a varying magnetic field, a voltage is induced across it. The level of the voltage is directly proportional to the number of turns and to the speed at which the magnetic field varies. Note that it does not matter whether coil is stationary in

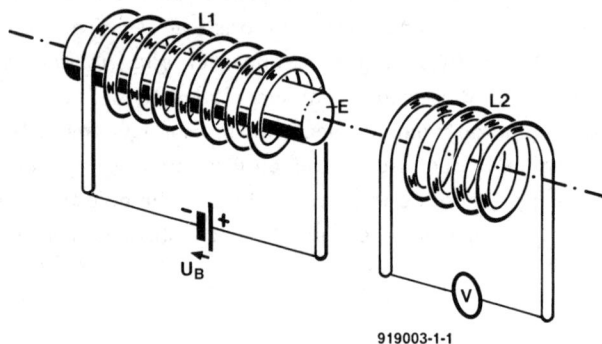

919003-1-1

Figure 1.1.1. Pinciple of voltage generation by means of induction.
When iron core E is moved, the magnetic field produced by coil L_1 changes
which causes a potential to be induced across coil L_2.

the varying magnetic field, or whether the coil moves through a magnetic field. The voltage, U, produced across the coil is

$$U = -n(d\phi/dt),$$
[Eq. 1b]

where n is the number of turns in the coil, ϕ is magnetic flux, and t is the time. The principle of generating such a voltage is shown in Figure 1-1. When a core, E, is inserted into coil L_1, the magnetic field caused by the current through L_1, and thus the density of the magnetic lines of force enclosed by L_2, changes. As long as the change lasts, a voltage is induced in L_2, whereupon voltmeter V gives a reading. When the core is withdrawn from coil L_1, the magnetic field decays, the number of lines of force enclosed by L_2 diminishes and the sign of the induced voltage changes (from + to − or from − to +, as the case may be), so that the voltmeter deflects to the other side.

The principle just described is frequently used in sound engineering, for instance, in dynamic microphones, pickup cartridges, and tape recorders.

Voltage generation by chemical reactions
When two conductors (electrodes) of dissimilar material are immersed in a weak solution of acid, alkaline or salt, a voltage ensues between them. The level of this voltage or potential is determined by the order of the conductors in the potential series of elements.

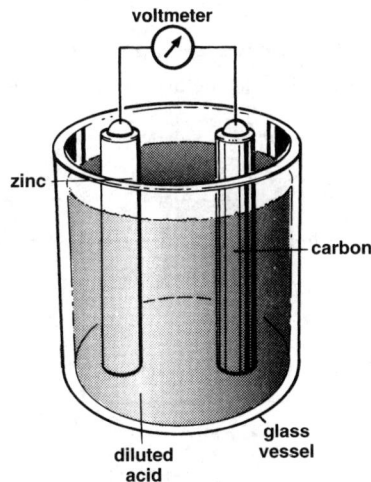

919003-1-2

Figure 1.1.2. Voltage generation through chemical reacion: when two dissimilar electrical conductors are placed in a conductive solution, a potential is generated between them.

3

The more the orders of the conductors are separated, the higher the voltage. This principle is shown in Figure 1.1.2.

When the electrodes are zinc and carbon, a voltage of 1.5 V is produced. In sound engineering, this principle is used in batteries, dry as well as rechargeable, for powering portable equipment. It is important for the user to know the life of the battery, since this is a measure of how long the portable equipment can be used before fresh batteries have to be inserted. Because of the importance of batteries, they will be reverted to in greater detail later in the book.

Voltage generation by mechanical pressure
When certain crystals, called piezo-electric crystals, are subjected to pressure, a voltage is produced at their surface—see Figure 1.1.3. This process is called the piezo-electric effect: it is used in, for instance, ceramic pick-up cartridges and crystal microphones.

Voltage generation by other means
Other means of generating a voltage are heat, light or friction, to name but a few. Since these processes are of no interest in sound engineering, they will not be discussed in this book.
A number of voltage sources and some of their properties are listed in Table 1.1.1.

1.1.2. Electric current

If there is a potential difference and a conducting path between two points, an electric current flows from one point to the other. In metals, an electric current consists entirely of

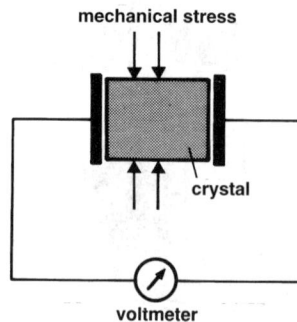

919003-1-3

Figure 1.1.3. Voltage generation by mechanical pressure.
When a certain crystal is distorted by pressure (strain or stress),
a potential is generated across its surface (piezo-electric effect).

Description	Manner of generation	Kind of voltage	Magnitude of voltage (V_{rms})
Output of electric guitar	inductive	alternating	about 0.1
Size HP7/11 dry battery	chemical	direct	1.5
Size PP3 dry battery	chemical	direct	9
Car battery	chemical	direct	12
Mains supply	inductive	alternating	240
Peak voltage in TV receiver	inductive	direct	24 000
High-voltage lines	inductive	alternating	380 000

Table 1.1.1.

a displacement of electrons from the point with a negative potential to the point with a positive potential as shown diagrammatically in Figure 1.1.4. The strength of the current is a measure of the number of electrons passing a cross-section of the conductor in unit time. The unit of current is the ampere, named after the French scientist Ampère. The symbol for current is I.

1.1.3. Direct current and alternating current

There is a fundamental difference between direct current and alternating current. Over a short time interval, the level of a direct current (or a direct voltage) does not alter. In other

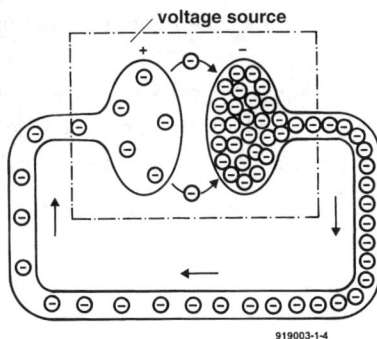

919003-1-4

Figure 1.1.4. In an electric current, negative charge carriers (electrons) move through a conductor. They move from the negative to the positive terminal of a voltage source.

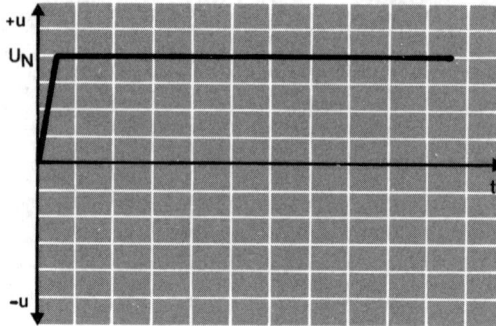

919003-1-5

Figure 1.1.5. Characteristic representing the function U=f(t) of a direct voltage. The magnitude of the voltage, U_N, is the same at any one time.

words, a direct current (or direct voltage) does not change as a function of time or, expressed mathematically,

$$I \quad f(t) \qquad \text{[Eq. 2a]}$$

and

$$U \quad f(t) \qquad \text{[Eq. 2b]}$$

The characteristic curve of a direct voltage, positive w.r.t. earth, vs time is shown in Figure 1.1.5. All electronic equipment, including that used in sound engineering, needs a direct voltage supply, irrespective of whether it is based on solid-state devices (transistors, diodes, integrated circuits) or thermionic valves (electron tubes). Normally, the direct voltage is derived by rectification of the (alternating) mains supply (US: household AC supply).

In contrast, an alternating voltage, or alternating current, does not remain constant over even a short time interval. If the alternating mains voltage is applied across a pure resistive load, the instantaneous values of voltage, u, and current, i, are given by

$$u = \hat{u}\sin(2\pi ft) \qquad \text{[Eq. 3a]}$$

and

$$i = \hat{\imath}\sin(2\pi ft) \qquad \text{[Eq. 3b]}$$

where $\hat{u}$ and $\hat{\imath}$ are the peak values of the alternating voltage and alternating current respectively, f is the frequency of the alternating voltage or current, t is the time. Often, ω is used for $2\pi ft$. The equations show that an alternating voltage or current varies with

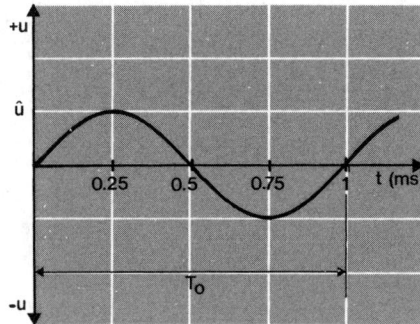

919003-1-6

Figure 1.1.6. Graph representing the function U=f(t) of an alternaing current. The waveform repeats itself every period T_0 (here 1 ms = 0.001 s)

time.

Apart from u and i, the period duration T is needed to describe an alternating voltage or current. This duration indicates how long a periodic phenomenon lasts until it repeats itself. The inverse of period duration is the frequency f:

$$f = 1/T \qquad\qquad\qquad [Eq.4]$$

The frequency indicates how many times per second a complete period, or cycle, is repeated. The unit of frequency is the hertz, abbreviated to Hz. The variation of an alternating voltage with a frequency of 1000 Hz as a function of time is shown in Figure 1.1.6.

Since the level of an alternating voltage or current varies continuously, its root-mean-square (r.m.s.) value is normally used in practice. This is the value that a direct voltage or current must have to produce the same power. Expressed mathematically,

$$U_{rms} = \sqrt{\frac{1}{T}\int_0^T u^2 dt} \qquad\qquad\qquad [Eq.5a]$$

$$I_{rms} = \sqrt{\frac{1}{T}\int_0^T i^2 dt} \qquad\qquad\qquad [Eq.5b]$$

In the case of sinusoidal voltages and currents:

7

$U_{rms} = \hat{u}/\sqrt{2}$ [Eq. 6a]

and

$I_{rms} = \hat{\imath}/\sqrt{2}$ [Eq. 6b]

For instance, in the United Kingdom, the mains voltage (household AC supply) has an r.m.s. value of 230 V and a frequency of 50 Hz. The peak value calculated with Eq. 6a is:

$\hat{u} = U\sqrt{2} = 230 \times 1.414 = 325$ V.

The period duration is calculated with Eq. 4:

$T = 1/f = 1/50 = 20$ ms.

Virtually all audio-frequency voltages encountered in sound engineering, such as the output signal of an amplifier, have no direct-voltage component. At the same time, these voltages are hardly ever sinusoidal. Their frequency depends on the tones contained in the music or voice, their peak value on the sound level, and their timbre on the shape of the waveform.

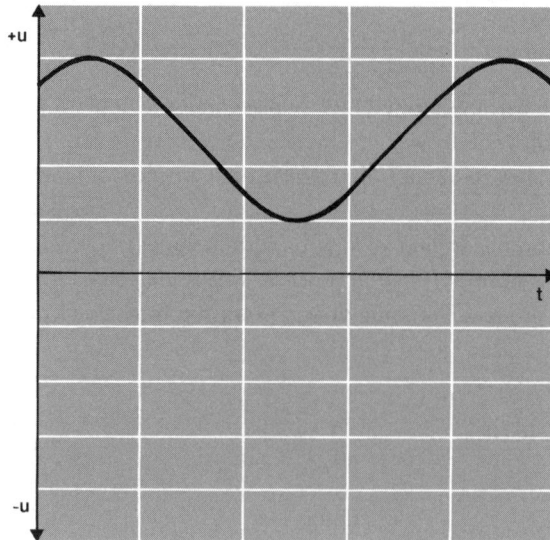

919003-1-7

Figure 1.1.7. Characteristic curve of an alternating (sinusoidal) voltage superimposed on to a direct voltage.

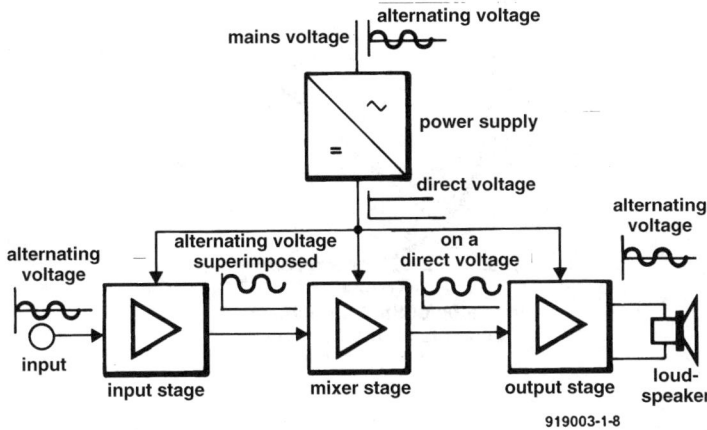

Figure 1.1.8. Simplified block diagram of an audio amplifier with indications as to where the voltage is alternating, direct or alternating superimposed on a direct voltage.

There are many instances where voltages and currents have an alternating as well as a direct component, such as the voltage whose waveform is shown in Figure 1.1.7. This is an alternating voltage superimposed on to a direct voltage. The block diagram of a guitar amplifier in figure 1.1.8. shows where in the circuit direct voltages, alternating voltages, or a combination of them may occur.

1.1.4. Electrical resistance and reactance

The strength of a current produced by a given electrical potential depends on the material and dimensions of the conductor. The behaviour of this conductor is described by its electrical resistance. The larger this is, the smaller the current; in other words, the current is inversely proportional to the resistance. The unit of electrical resistance is the ohm (after the mathematician Georg Ohm), normally indicated by the Greek letter Ω (omega); its symbol is R.

A distinction must be made between a pure resistance, whose value is constant and independent of the applied voltage, and a reactance, which is the electrical resistance offered by capacitors and inductors, and whose value depends on the frequency of the applied voltage. The reactance curves of an inductor and capacitor are shown in Figure 1.1.9. The reactance of a capacitor is indicated by the symbol X_C and that of an inductor by X_L. Reactance, like resistance, is measured in ohms.

$$X_C = 1/2\pi fC, \text{ or (since } 2\pi f = \omega), = 1/\omega C, \qquad\qquad \text{[Eq. 7]}$$

9

**the value of the LC impedance
depends on the frequency**
919003-1-9

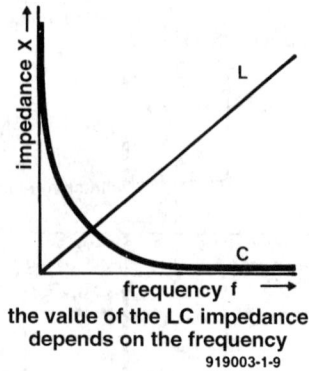

*Figure 1.1.9. Characteristic curve of the reactance X = f(f) of an
inductance L and a capacitor C. The reactance of L is directly proportional to the
frequency, whereas that of the capacitor is inversely proportional to f.*

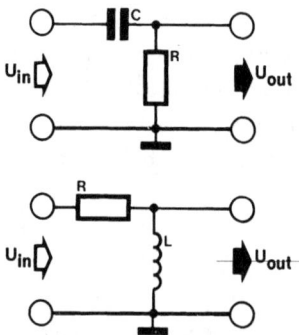

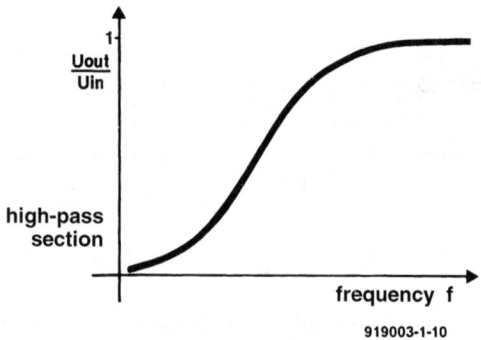

919003-1-10

*Figure 1.1.10. Circuit and frequency response curve [U$_{out}$/U$_{in}$=f(f)] of a high-pass
filter. Low frequencies are blocked, while high frequencies are passed.*

and

$$X_L = \omega L \qquad\qquad\qquad \text{[Eq. 8]}$$

where L is the self-inductance of the coil in henry (H).

In electronic systems, resisance and reactance normally occur side by side. The sum of their total resistance is called the impedance, symbol Z.

Ingenious combinations of resistors, capacitors and inductors may create networks

10

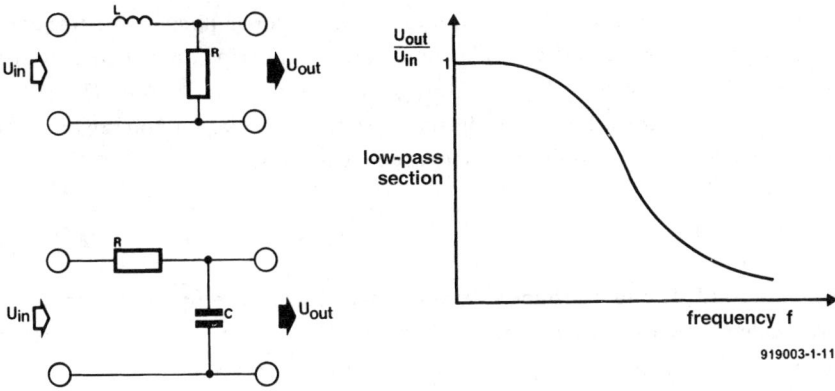

Figure 1.1.11. Circuit and frequency response curve $[U_{out}/U_{in}=f(f)]$ of a low-pass filter. Low frequencies are passed, while high frequencies are blocked.

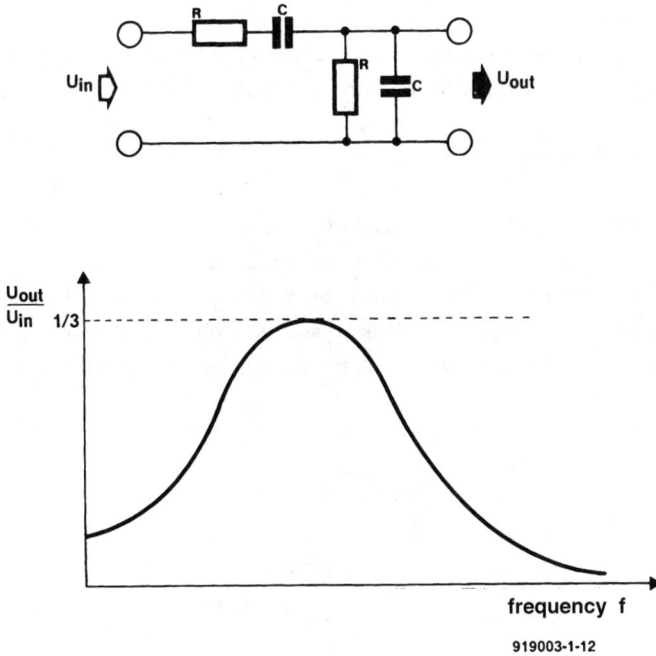

Figure 1.1.12. Circuit and frequency response curve $[U_{out}/U_{in}=f(f)]$ of a simple bandpass filter. Both low and high frequencies are blocked, while centre frequencies are passed.

11

that by means of frequency-dependent voltage division can influence the frequency response of a signal. The result is called linear distortion. The simplest examples of such networks are the high-pass filter (Figure 1.1.10), the low-pass filter (Figure 1.1.11), and the band-pass filter (Figure 1.1.12). Such filters are used in a variety of applications, from tone control circuits to cross-over filters in loudspeaker systems.

1.1.5 Ohm's Law

Voltage, U, current, I, and resistance, R, are the three most important quantities in electronics technology. They are related by Ohm's Law, which, in simplified form, states that

$$U = IR \qquad\qquad \text{[Eq. 9a]}$$

$$R = U/I \qquad\qquad \text{[Eq. 9b]}$$

$$I = U/R \qquad\qquad \text{[Eq. 9c]}$$

Although, in practical terms, Ohm's Law is rather more complex, these simplified equations are perfectly adequate for a good understanding of the material in this book.

1.1.6 Electrical power

When we switch on a supply, we expect that something useful will happen, such as the amplification of a microphone signal; in other words, we expect power to be delivered. The electrical power that can be delivered by a supply is calculated by the quantities voltage and current. Electrical power is expressed in watts (W) after the famous English physicist, James Watt. Its symbol is P. Electrical power can be expressed by various equations:

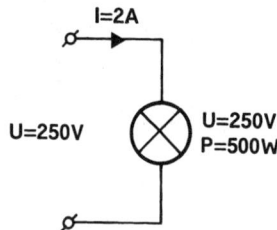

Figure 1.1.13. Circuit symbol of an incandescent bulb with associated current and voltage.

$$P = UI = U^2/R = I^2R \qquad\qquad\qquad \text{[Eq. 10]}$$

Electrical powers used in sound engineering are small in comparison to those used in industrial power supplies or in heavy engineering. The acoustical power emanating from loudspeakers is smaller still. This will be made clearer by two examples.

The lamp in the circuit in Figure 1.1.13, carries the following data: 250 V (which is its voltage rating), and 500 W (which is its power rating, P). When a voltage of 250 V is applied to the lamp, the current flowing through it is according to Eq. 10:

$$I = P/U = 500/250 = 2 \text{ A.}$$

The resistance of the lamp is according to Eq. 9:

$$R = U/I = 250/2 = 125 \ \Omega.$$

When the applied voltage is lowered to 230 V (Figure 1.1.14), the power according to Eq. 10 is

$$P = U^2/R = 230^2/125 = 423.2 \text{ W.}$$

This reduction in power will be noticeable through the reduced light coming from the lamp.

The second example refers to a domestic mains installation, which is normally divided into different groups, each of which has its own fuse or automatic cutout. A group may contain a number of socket outlets or several lights. In the simplest case, a group has only one socket outlet or only one socket outlet is used. In the case of a group with a signal socket outlet that is protected by a 20 A fuse, the maximum power available from the outlet is

$$P = UI = 230 \times 20 = 4600 \text{ W.}$$

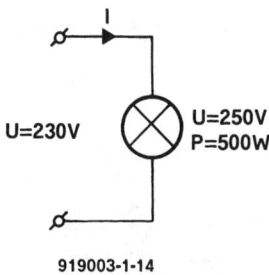

U=230V U=250V P=500W

919003-1-14

Figure 1.1.14. The same bulb as shown in Figure 1.1.13, but now connected to an alternating voltage of 230 V.

When more power is taken from the outlet, the fuse will blow or the cut-out will trip. The maximum number of lamps that can be connected to the outlet simultaneously is

4600/423.2 = 10.87 or, rounded down, 10.

It should be borne in mind that each electrical device draws a much larger current immediately after it has been switched on than when it has settled in: the switch-on current is many times larger than the normal operating current. It is, therefore, essential that the eight lamps are not switched on at the same time, but one by one or in groups of two. If they were switched on simultaneously, the fuse would undoubtedly blow or the cut-out trip. In general: equipment that draws a large current, such as public-address amplifiers, should not be switched on at the same time, but one after another.

1.1.7 Efficiency

Electrical equipment converts energy from one shape to another. Applied electrical energy, for instance, may be converted into sound energy or light. It is impossible for the applied energy to be completely converted into another kind of useful energy. For example, in the conversion from electrical energy into sound energy, part of the applied energy is lost. The applied energy, W_{in} is always larger than the useful energy, W_{out}. The ratio of useful energy to applied energy is called efficiency, symbol η, so that

$$\eta = W_{out}/W_{in} = P_{out}/P_{in} = <1 \qquad \text{[Eq. 11]}$$

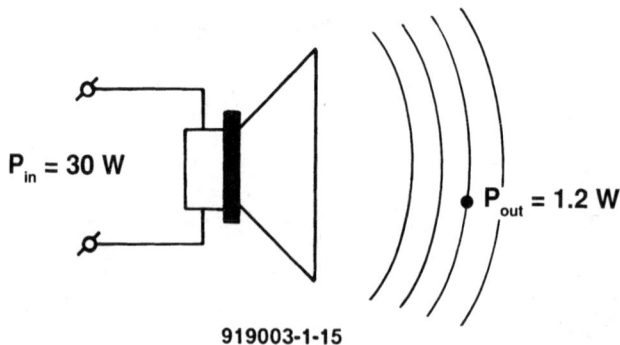

919003-1-15

Figure 1.1.15. Circuit symbol of a loudspeaker to which an electrical power P_{in}=30 W is applied and which radiates a sound power of P_{out}=1.2 W.

14

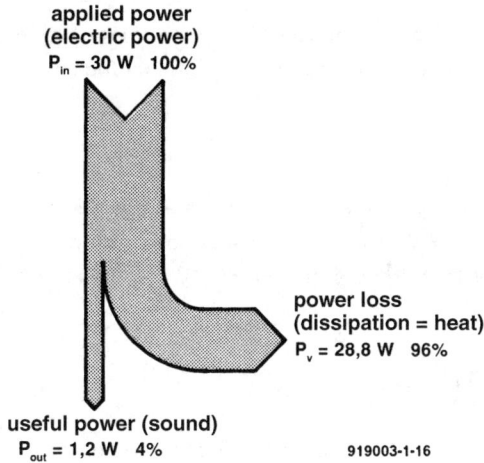

**applied power
(electric power)**
$P_{in} = 30\ W\quad 100\%$

**power loss
(dissipation = heat)**
$P_v = 28,8\ W\quad 96\%$

useful power (sound)
$P_{out} = 1,2\ W\quad 4\%$

919003-1-16

*Figure 1.1.16. Schematic representation of the power distribution
with the loudspeaker shown in Figure 1.1.15.*

η is often expressed as a percentage, for which purpose it is simply multiplied by 100. If, for example, audio power of 30 W is applied to a loudspeaker (see Figure 1.1.15) and the radiated sound power is about 1.2 W, the efficiency of the loudspeaker is

$\eta = 1.2/30 = 0.04$ (or, $0.04 \times 100 = 4\%$).

This means that 96% of the electrical power is converted by the loudspeaker into an unwanted form of energy (in this case: heat—see Figure 1.1.16). Note that this is not an unusual value for a low-frequency loudspeaker.

1.1.8 Batteries

Batteries are playing a more and more important role in sound engineering since musicians increasingly want to be free to move about without having to rely on mains power. Battery-operated instruments and equipment provide this freedom. However, some aspects of these chemical sources of energy must be watched. One of the most important, if not *the* most important, of these is that we need to know how long a battery is able to power a particular instrument or piece of equipment. It does not look very professional if an instrument fails owing to lack of power during a live performance. Even if the battery or batteries can be replaced during a natural interval, it creates a stress situation for the musician(s) that should have been prevented. To estimate the life, t, of a

15

battery, we need to know its capacity C, and the average current, I, drawn by the relevant instrument or equipment. The (theoretical) life of the battery is given by

$$t = C/I, \hspace{8cm} \text{[Eq. 12]}$$

where C is in ampere-hours (Ah), I is in amperes (A), and t is in hours.

If, for instance, a 9 V battery of medium quality with a capacity of about 0.2 Ah is used to power a mixer that draws a current of 0.01 A, it will be able to deliver energy for a period of

0.2/0.01 = 20 hours.

Usually, the life so estimated is too high. This may be because at the start the battery is not fully charged, or, when the battery is discharged to 80% of its nominal capacity, its terminal voltage drops to a level that is too low for the relevant instrument or equipment. All this means that there is no guarantee that the estimated life will be realized. The capacity and nominal e.m.f. of some frequently used dry alkaline-manganese batteries are given in Table 1.1.2. Note that the values are nominal ones which may vary from manufacturer to manufacturer.

The current drawn by instruments and equipment must be obtained from their published technical data, but may also be obtained readily by measuring it, which is best doen as shown diagrammatically in Figures 1.1.17 and 1.1.18. The multimeter should be set to a direct-current (d.c.) range, always starting with the highest. It is, of course, necessary to observe correct polarity and to avoid short-circuits. Almost any type of multimeter may be used as long as it has a d.c. range. The equipment in the photograph in Figure 1.1.18 draws a current of 13.8 mA.

Some equipment starts to draw a current automatically as soon as the cable from the

Type of battery	Nominal e.m.f.	Nominal capacity
N, AM5, LR1, MN9100	1.5 V	0.5 Ah
AAA, AM4, MN2400, HP16	1.5	0.8 Ah
AA, AM3, RG, HP7, MN1500	1.5 V	1 Ah
C, AM2, R14, HP11, MN1400	1.5 V	1.5 Ah
D, AM1, R20, HP2, MN1300	1.5 V	6 Ah
PP3, 6AM6, 6F22, MN1604	9.0 V	0.25 Ah

Table 1.1.2

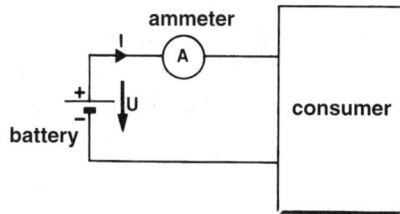

919003-1-17

Figure 1.1.17. Circuit for measuring the current drawn from a battery
by a consumer (equipment or instrument).

power source is connected to it, but only draws its full operating current when it is
switched on properly. This will be reverted to later when the use of test equipment is
discussed.

Experienced sound engineers are able to estimate the current drawn by a particular

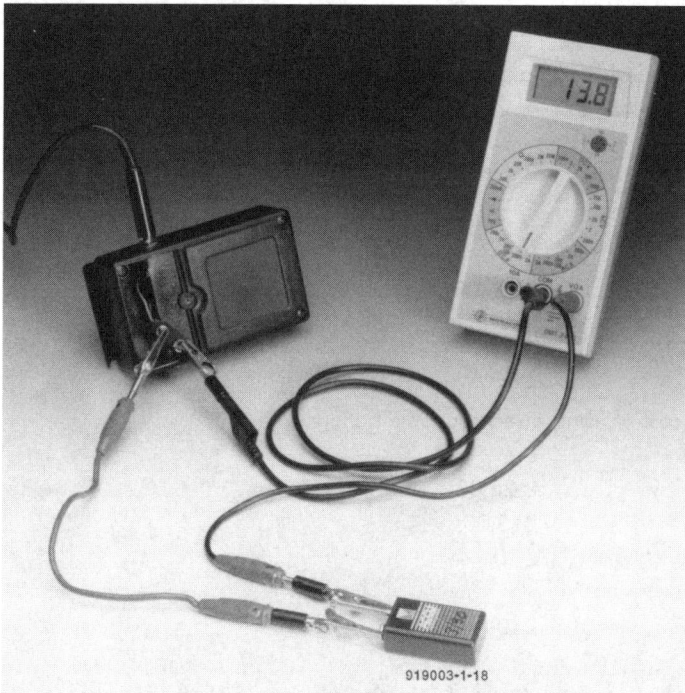

919003-1-18

Figure 1.1.18. Actual setup for measuring the current drawn from a battery
by a consumer (equipment or instrument).

17

instrument or equipment fairly accurately. Owing to the increasing use of portable equipment and instruments, the cost of batteries keeps on rising and it is, therefore, sensible to consider alternatives to dry alkaline manganese batteries, such as rechargeable NiCd (nickel-cadmium), Li-ion (lithium ion) or NiMH (nickle metal hydride) batteries.

A cost comparison of, for instance, a dry and a rechargeable (NiCd) 9 V battery would look something like this. A single dry alkaline-manganese 9 V battery costs about £2 and a rechargeable NiCd one around £25 (nickel metal hydride £5; lithium manganese £6) . With careful use the number of charge-discharge cycles of the NiCd battery is 500–1000. Normally, a dry battery has twice the capacity of a comparable rechargeable one, that is, it lasts twice as long as a rechargeable one before this needs to be recharged. This means that a NiCd battery has a life of 250–500 times that of a dry battery; in other words, a NiCd battery is equivalent to that number of dry batteries. So, operating a unit from dry batteries would cost £500–£1000.

The energy released when a 9 V, 110 mAh NiCd battery is discharged amounts to

$$110 \times 10^{-3} \times 9 = 0.99 \text{ Wh.}$$

The energy required to charge a flat 9 V, 100 mAh NiCd battery is at worst twice the released energy, that is

$$2 \times 0.99 = 1.98 \text{ Wh.}$$

Assuming that electricity costs 7 pence per kWh, the total costs of charging the battery 1000 times amount to

$$1000 \times 1.98 \times 10^{-3} \times 0.07 \approx 14 \text{ pence.}$$

So, the total cost of using a rechargeable battery is

NiCd battery	£ 25·00
Charger	£15·00
Electricity	£ 0·14
	£40·14

This example shows that the cost of using a rechargeable battery is about 4–8 per cent of that when a dry battery is used. However, it should be borne in mind that a NiCd battery has only half the capacity of a comparable dry battery and that its self-discharge is greater than that of a dry battery. Therefore, its use is only fully justified in equipment that is used frequently and that does not draw too high a current, such as guitar effects units, portable keyboards and mixers. For equipment that draws relatively large currents, such as portable tape recorders, the capacity of NiCd batteries is normally not sufficient to guarantee reasonably long operation between two charging cycles. On the other hand, in equipment that draws only a very small current, such as capacitor microphones, the battery is usually exhausted through self-discharge. In such cases, it is, therefore, wiser to use dry batteries.

When rechargeable batteries are used, the manufacturers' user instructions must be followed carefully to ensure as long a life of the battery as possible. Also, in the case of sintered NiCd batteries, the memory effect must be borne in mind. The nickel-cadmium battery is mechanically rugged and long-lived. It has excellent low-temperature characteristics and can be hermetically sealed. Cost, however, is higher than for the lead-acid or nickel-zinc battery. In many applications, the use of a sealed lead-acid battery is to be preferred over the other two types.

A slight drawback of a sintered NiCd battery is its so-called memory effect, which is fortunately completely reversible. It should be noted that mass plate nickel-cadmium cells and batteries do not develop the memory effect in any circumstances.

Low internal resistance
The ability of NiCd batteries to provide fairly large currents (because of their low internal resistance – at least as far as sintered types are concerned) is an important factor for the amateur fraternity, since many home-made model units draw fairly large currents. As a comparison, the d.c. resistance of three types of fully charged 1 Ah, 1.2 V sealed cell is

919003-1-19

Figure 1.1.19. From left to right: an AA size dry battery; a PP3 dry battery; an AA size rechargeable battery; and a rechargeable PP3 battery

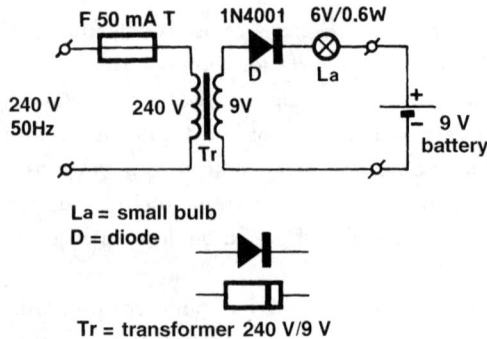

Figure 1.1.20. Circuit of a simple charger for 9 V batteries.
The small bulb provides primitive current regulation.

Standard	110 mΩ/cell
Heavy duty	50 mΩ/cell
Sintered	19 mΩ/cell

Environmental effects
One of the most serious drawbacks of NiCd batteries is their effect on the environment. This type of battery contains cadmium which is toxic. In most countries, discarded NiCd batteries are dumped on the rubbish heap where they remain toxic for a very long time. It is true, of course, that their life of some 500–800 charge/discharge cycles does not cause millions of them to be disposed of on the rubbish heap. Nevertheless, this was a very important factor in the decision of manufacturers in general to discontinue the use of NiCd batteries in most consumer products.

Memory effect
Another disadvantage of sintered (not mass plate) NiCd batteries is, as already mentioned, their memory effect. This manifests itself in the cell retaining the characteristics of previous cycling. That is, after repeated shallow-depth discharges the cell will fail to provide a satisfactory full-depth discharge. Note, however, that Eveready cylindrical nickel-cadmium cells are particularly noted for their lack of memory effect.

The memory effect is a nuisance, because it means that a battery with a nominal capacity of, say, 600 mAh, after a number of charge/discharge cycles has a useful capacity of only 300 or 400 mAh. This has nothing to do with the life of the battery: even a new battery if charged as stated will soon lose part of its capacity.

Fortunately, this reduction in capacity can be prevented fairly simply. Moreover, batteries that already suffer from the memory effect can be restored to their nominal capacity. The cure is simply to ensure that a battery is occasionally fully discharged before it is recharged. Occasionally means before every third or so recharge. Note that there are chargers on the market that have the discharge facility built in, but this will certainly not be the case in the less expensive types.

Correct discharging
There is no need for extensive circuitry to discharge a battery: a simple resistor or light bulb will accomplish it readily. It is, however, necessary to keep an eye on the discharge time, because otherwise there is the risk that the battery is discharged beyond a certain voltage. When this happens, it may cause polarity reversal in the cells comprising the battery.

   NiCd batteries must always be charged by appropriate chargers. The complete discharge of, and also overcharging, a NiCd battery may lead to irreversible degradation of its capacity. The charging current, I_c, during a normal charging cycle must be

$$I_c = C/10$$

For example, the charging current for a 9 V, 110 mAh NiCd battery must not exceed 0.11/10 = 0.011 A = 11 mA. In the case of a fully discharged battery, this charging current will recharge the battery in about 14 hours. This period must not be be exceeded by more than an hour. A simple charger for 9 V NiCd batteries may be constructed as shown in Figure 1.1.20.

1.1.9 Resistor networks

All electrical users, such as loudspeakers, lamps, and amplifiers, have electrical resistance R, and may, therefore, be considered as resistors. Resistors may be combinaed in various ways, with varying results. Certainly, when a loudspeaker is to be linked to an amplifier, it is of vital importance to be able to anticipate what the result will be: this knowledge may prevent a lot of damage.

   When resistors are connected in series, the total resistance is the sum of their individual resistances. The total resistance may therefore be represented by a single resistor as shown in Figure 1.1.21. The sum of the potential differences across the resistances, U_1, U_2, U_3, ... is equal to the total potential difference, U_t, across the series network. These properties are expressed by the following equations.

$$R_t = R_1 + R_2 + R_3 + \ldots + R_n \qquad \text{[Eq. 13]}$$

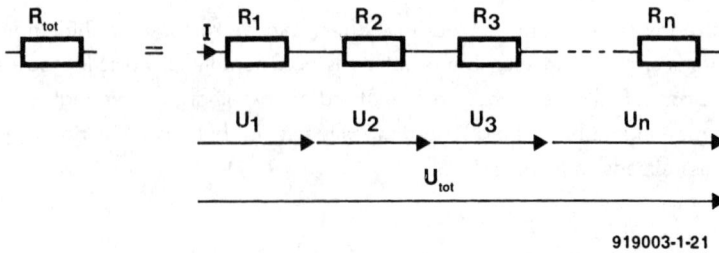

919003-1-21

Figure 1.1.21. Resistors connected in series: the total resistance
is equal to the sum of the individual resistors.

$$U_t = U_1 + U_2 + U_3 \ldots + U_n \qquad \text{[Eq. 14]}$$

In a series network of resistors, the current is the same through all resistors.

$$I = U_1/R_1 = U_2/R_2 = U_3/R_3 = U_n/R_n$$

For example, when two loudspeakers are connected in series as in Figure 1.1.22, and the resistance of one is 8 Ω and that of the other is 4 Ω, the total resistance is 12 Ω. The amplifier to which the series combination is linked, delivers a power of 30 W into 12 Ω. This means that the amplifier and loudspeaker network are well matched. The power across each of the loudspeakers is calculated as follows.

$$P_t = I^2R = P_1 + P_2$$

$$\therefore I^2 = (P_1 + P_2)/R, \text{ so that } I = \sqrt{[(P_1 + P_2)/2]}$$
$$I = \sqrt{(P_1/R_1)} = \sqrt{(P_1/R_1)} = \sqrt{[(P_t - P_2)/R_1]}$$

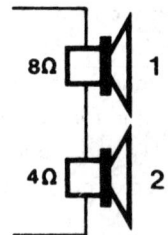

Figure 1.1.22. Two loudspeakers of unequal input impedance
connected in series.

919003-1-22

22

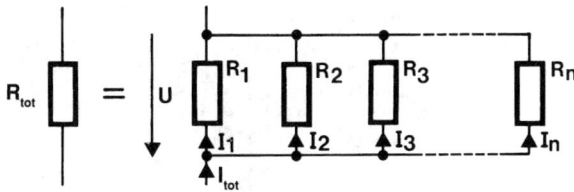

Figure 1.1.23. Resistors connected in parallel: the total resistance
is smaller than that of the lowest-value individual resistor.

$\therefore P_2 = P_t/[(R_1/R_2)+1] = 30/[8/4)+1] = 10$ (W)

so that the power, P_1, dissipated in the loudspeaker is

$P_1 = 30-10 = 20$ W

When resistors are in parallel, the reverse of the total resistance is equal to the sum of the
reverse of the individual resistors (Figure 1.1.23).

$1/R_t = 1/R_1 + 1/R_2 + 1/R_3 + \ldots + 1/R_n$ [Eq. 15]

(It should be noted that the reverse of resistance is conductance, symbol G, which is
expressed in siemens, S. So, a resistor, R, of 10 Ω has a conductance, G, of 0.1 S).
The sum of the currents, I_t, through the individual resistors, $I_1, I_1, I_1 \ldots I_1$ is

$I_1 = I_1 + I_2 + I_3 + \ldots + I_n$ [Eq. 16]

In a parallel network of resistors, the potential across the network is equal to that across
each individual resistor.
Example. The two loudspeakers from the previous example, impedance 8 Ω and 4 Ω
respectively, are connected in parallel and linked to an amplifier that has a rated output
impedance of 4 Ω – see Figure 1.1.24. According to Eq. 15, the total resistance, R_t, is

$R_t = 1/(1/R_1 + 1/R_2) = R_1R_2/(R_1+R_2) = 32/12 = 2.67$ Ω

With such a relatively low load, the amplifier would probably be overloaded; in any case,
the amplifier and loudspeaker combination are not matched To obtain a correct match, a
resistor, R_s, of 4 Ω must be connected in series with loudspeaker 2 – this results in the

23

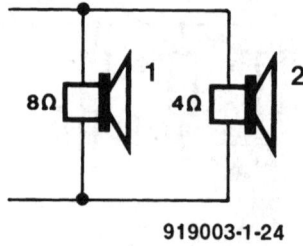

919003-1-24

Figure 1.1.24. Two loudspeakers of unequal impedance connected in parallel.

circuit in Figure 1.1.25.

The total resistance of the network is then

$$R_t = R_1 R_2/(R_1 + R_2) = 64/16 = 4 \ \Omega.$$

It is necessary to determine the power dissipated in the resistor. Since the resistance of the network is 4 Ω and the amplifier output is 30 W, according to Eq. 10:

$$P = U^2/R_t,$$

so that

$$U = \sqrt{(P/R_t)} = \sqrt{(30 \times 4)} \approx 11 \ V.$$

The current through the branch in which the series resistor is located is

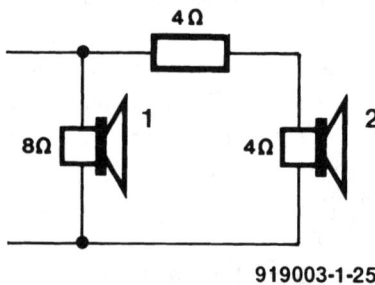

919003-1-25

Figure 1.1.25. The 4 Ω resistor ensures correct matching to the amplifier output.

$$I = U/R_2 = 11/8 = 1.38 \text{ A}.$$

Therefore, the power dissipated in, that is, the rating of, the resistor, P_R is

$$P_R = I^2 R_s = 1.38^2 \times 4 = 7.6 \text{ W}.$$

A special case of a series network of resistors is the potential or resistive divider. Ready-made potential dividers are commercially available under the name potentiometer. They are normally formed from carbon resistor track across which a wiper can be moved. In principle, there are two different constructions: rotary and slide potentiometers – see Figure 1.1.26, which gives two alternative circuit symbols for the device. Depending on the position of the wiper, the output voltage, U_o can have any value between 0 V and the input voltage, U_i. When the wiper is exactly at the centre of the track, in the case of a linear potentiometer, $U_o = U_i/2$.

Potentiometers are encountered in virtually all sound equipment for most diverse functions. The simplest and best known is that of volume control in an amplifier. When the sound level of an amplifier needs to be increased, the volume control is adjusted so that a larger part of the audio frequency input voltage is taken from the potential divider and applied to the amplifier.

If a potentiometer cannot be operated by hand, but only by a suitable screwdriver, it is

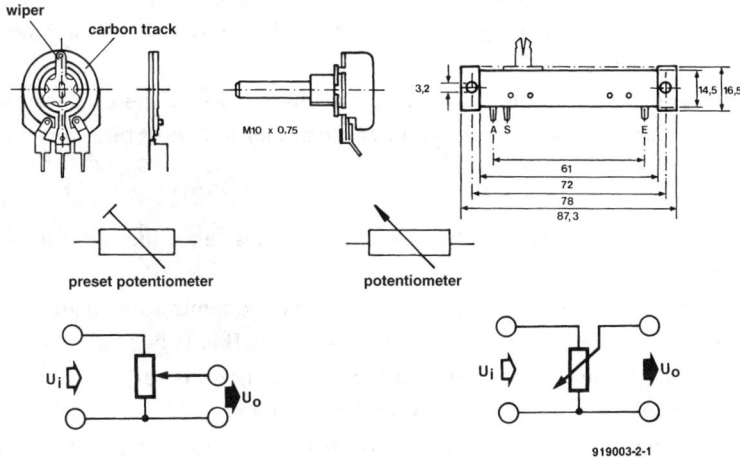

Figure 1.1.26. Circuit symbols and practical construction of some typical potentiometers, including a preset model (without spindle). Potentiometers may be rotary types or slide types (as often used in mixers).

25

called a preset potentiometer (normally abbreviated to preset). Presets are used in electronic circuits when they need to be set only once (normally by a test engineer or technician).

There are also presets with a non-linear, normally a logarithmic, characteristic. In these, the resistance does not increase in linear, but in logarithmic, proportion to the angle of rotation of the device. From 0 Ω, the resistance initially changes little, but as the preset is turned open further and further, the resistance increases faster and faster. In Figure 1.1.27, the x-axis represents the angle of rotation in degrees, while the y-axis represents the output voltage as a percentage of the input voltage.

1.1.10 Measurement techniques

In the foregoing, much has been said about the voltage, U, the current, I, and the resistance, R, but in practice the value of these quantities is often not known. If the values cannot be estimated from experience, or when an equipment is faulty, they will have to be measured. The most suitable instrument for this is a sturdy, digital multimeter like the one shown in Figure 1.1.28. The result of the measurement is shown in ciphers, which reduces the likelihood of error to a minimum. When buying one, make sure that it is sturdy and lies easily in the hand. Sensitive analogue instruments are less suitable for the often rough treatment it may receive within a band. Moreover, reading an analogue instrument is not always easy for an amateur (although even professionals have been known to make mistakes). On the other hand, digital instruments, especially the less expensive ones, have a measurement bandwidth of only 400–500 Hz so that they are not suitabble for measuring audio frequency signals.

As regards the operation of the instrument, it is best to refer to the user instructions supplied with the meter. However, there are three strict rules that must be observed at all times.

1. Whenever current or voltage is to be measured, always select the highest relevant range to start with.
2. When a current or resistance range is selected, never attempt to measure voltage, since this will almost certainly destroy the instrument. This is because the internal resistance of the instrument when a current range is selected is very small, so that at almost any voltage a very high (destructive) current will flow.
3. Make it a habit whenever measurements have been carried out to select the highest voltage range before putting the instrument away (it must, of course, also be switched off).

Measuring a voltage is a fairly simple affair. Start by selecting the highest range for the

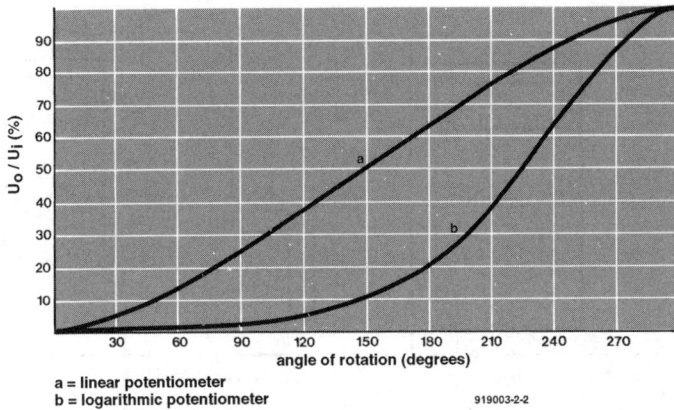

a = linear potentiometer
b = logarithmic potentiometer

919003-2-2

Figure 1.1.27. Typical characteristic curves of two popular potentiometers.

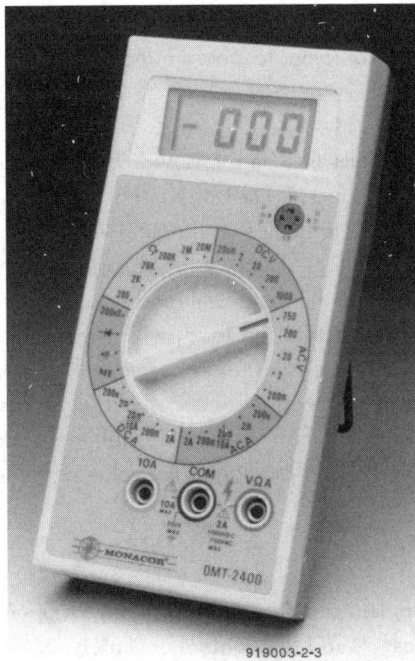

919003-2-3

Figure 1.1.28. Standard multimeter for measuring direct voltages (DCV), alternating voltage (ACV), direct currents (DCA), alternating currents (ACA) and resistance. The meter shown is set to the 750 V ACV range.

27

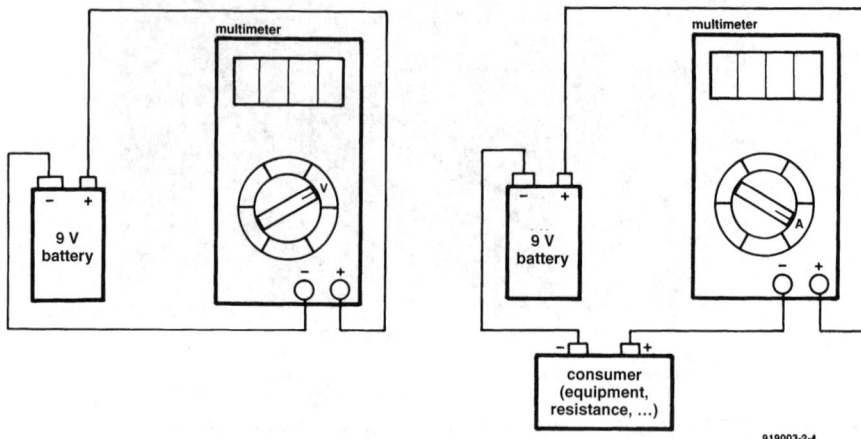

Figure 1.1.29. Correct ways of using a multimeter. On the left it is connected in parallel with a battery to measure its (open-circuit) voltage, and on the right it is connected in series with the battery and consumer to measure the current drawn from the battery.

type of voltage (direct or alternating) to be measured. Then, connect the meter in parallel with the voltage source—see Figure 1.1.29. If it is not known whether the voltage is a direct or alternating one, select the highest alternating voltage range. If the meter reads nothing, try the highest direct voltage range. If the voltage consists of an alternating one superimposed on a direct voltage, the meter will read something in either range.

Current is measured in series with the load, which means that the circuit must be broken to insert the meter. The same rules apply as with voltage measurements.

Resistance must be measured only if there is no voltage across it. The resistance of a loudspeaker can be measured by connecting the multimeter across the loudspeaker terminals. The meter will give a direct reading of the value of the resistance.

1.1.11 Safety

In all mains-operated equipment certain important safety requirements must be met. The relevant standard for most sound equipment is *Safety of Information Technology Equipment, including Electrical Business Equipment* (European Harmonized British Standard BS EN 60950:1992. Electrical safety under this standard relates to protection from

- a hazardous voltage, that is, a voltage greater than 42.4 V peak or 60 V d.c.;
- a hazardous energy level, which is defined as a stored energy level of 20 Joules or

28

more or an available continuous power level of 240 VA or more at a potential of 2 V or more;

a single insulation fault which would cause a conductive part to become hazardous;

- the source of a hazardous voltage or energy level from primary power;
- secondary power (derived from internal circuitry which is supplied and isolated from any power source, including d.c.)

Protection against electric shock is achieved by two classes of equipment.

Class I equipment uses basic insulation ; its conductive parts, which may become hazardous if this insulation fails, must be connected to the supply protective earth.

Class II equipment uses double or reinforced insulation for use where there is no provision for supply protective earth (rare in sound engineering – mainly applicable to power tools).

If Class II equipment is totally enclosed by non-conductive durable insulating material , it is called 'insulation-encased Class II equipment'. If it is totally enclosed by a metallic enclosure in which double or reinforced insulation is used throughout, it is called 'metal-encased Class II equipment'. Even so, where supply protective earth is available, it should be used if at all possible.

The use of a a Class II insulated transformer is preferred, but note that when this is fitted in a Class I equipment, this does not, by itself, confer Class II status on the equipment.

Electrically conductive enclosures that are used to isolate and protect a hazardous supply voltage or energy level from user access must be protectively earthed regardless of whether the mains transformer is Class I or Class II.

There is no requirement for a safety earth if the hazardous supply voltage or energy level area enclosure is non-conductive to Class II insulation requirements and the insulation in the mains transformer is Class II, although a mains earth may still need to be connected for functional purposes.

Always keep the distance between mains-carrying parts and other parts as large as possible, but never less than required.

If at all possible, use an approved mains entry with integrated fuse holder and on/off switch. If this is not available, use a strain relief (Figure 1.1.30, note 2) on the mains cable at the point of entry. In this case, the mains fuse or circuit breaker should be placed after the double-pole on/off switch unless the fuse is a Touchproof® type or similar. Close to each and every fuse must be affixed a label stating the fuse rating and type. The rating of a slow fuse should be not greater than 1.25 times the normal operating current, whereas that of a fast fuse should be equal to the operating current. Fast fuses are used, for instance, in case of multiple secondary windings, but if there is an electrolytic capacitor behind the secondary, a slow fuse must be used to allow for surges in the charging

1. Use a mains cable with moulded-on plug.
2. Use a strain relief on the mains cable.
3. Affix a label at the outside of the enclosure near the mains entry stating the equipment type, the mains voltage or voltage range, the frequency or frequency range, and the current drain or curent drain range.
4. Use an approved double-pole on/off switch, which is effectively the 'disconnect device'.
5. Push wires through eyelets before soldering them in place.
6. Use insulating sleeves for extra protection.
7. The distance between transformer terminals and core and other parts must be 6 mm.
8. Use the correct type, size and current-carrying capacity of cables and wires.
9. A printed-circuit board like all other parts should be well secured. All joints and connections should be well made and soldered neatly so that they are mechanically and electrically sound. Never solder mains-carrying wires directly to the board: use solder tags. The use of crimp-on tags is also good practice.
10. Even when a Class II transformer is used, it remains the on/off switch whose function it is to isolate a hazardous voltage (i.e., mains input) from the primary circuit in the equipment. The primary-to-secondary isolation of the transformer does not and can not perform this function.

Figure 1.1.30

Figure 1.1.31. Equipment provided with the mains Euro plug shown must be double insulated. The absence of an earth connection means that earth loops can not occur, but it may cause problems when an operational earth is required.

current.

The separate on/off switch (Figure 1.1.30, note 4), which is really a 'disconnect device', should be an approved double-pole type (to switch the phase and neutral conductors of a single-phase mains supply). In case of a three-phase supply, all phases and neutral (where used) must be switched simultaneously. A pluggable mains cable may be considered as a disconnect device. In an approved switch, the contact gap in the off position is not smaller than 3 mm.

The on/off switch must be fitted by as short a cable as possible to the mains entry

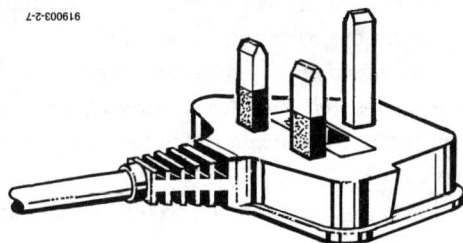

Figure 1.1.32. The standard mains plug (to BS1363A) in the United Kingdom is polarized and individually fused. Removal of the earth pin makes it impossible for the plug to be inserted into the mains outlet. Probably the safest mains plug in the world.

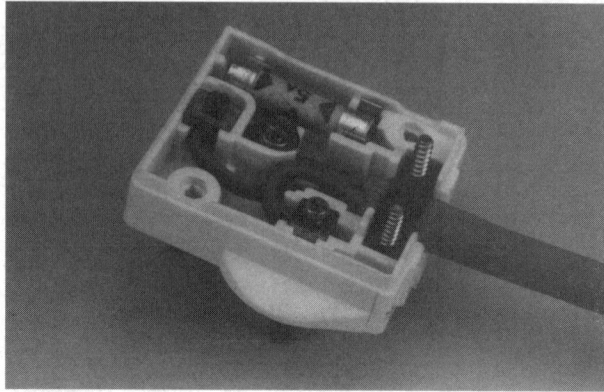

Figure 1.1.33. Correctly wired mains plug to BS1363A.
Note the screwed cable retainer and the individual fuse.

point. All components in the primary transformer circuit, including a separate mains fuse and separate mains filtering components, must be placed in the switched section of the primary circuit. Placing them before the on/off switch will leave them at a hazardous voltage level when the equipment is switched off.

If the equipment uses an open-construction power supply which is not separately protected by an earthed metal screen or insulated enclosure or otherwise guarded, all the conductive parts of the enclosure must be protectively earthed using green/yellow wire (green with a narrow yellow stripe – do not use yellow wire with a green stripe). The earth wire must not be daisy-chained from one part of the enclosure to another. Each conductive part must be protectively earthed by direct and separate wiring to the primary earth point which should be as close as possible to the mains connector or mains cable entry. This ensures that removal of the protective earth from a conductive part does not also remove the protective earth from other conductive parts.

Pay particular attention to the metal spindles of switches and potentiometers: if touchable, these must be protectively earthed. Note, however, that such components fitted with metal spindles and/or levers constructed to the the relevant British Standard fully meet all insulation requirements.

The temperature of touchable parts must not be so high as to cause injury or to create a fire risk.

Most risks can be eliminated by the use of correct fuses or circuit breakers, a sufficiently firm construction, correct choice and use of insulating materials and adequate cooling through heat sinks and by extractor fans.

The equipment must be sturdy: repeatedly dropping it on to a hard surface from a height of 50 mm must not cause damage. Greater impacts must not loosen the mains transformer, electrolytic capacitors and other important components.

Do not use dubious or flammable materials that emit poisonous gases.

Shorten screws that come too close to other components.

Keep mains-carrying parts and wires well away from ventilation holes, so that an intruding screwdriver or inward falling metal object cannot touch such parts.

As soon as you open an equipment, there are many potential dangers. Most of these can be eliminated by disconnecting the equipment from the mains before the unit is opened. But, since testing requires that it is plugged in again, it is good practice (and safe) to fit a residual current device (RCD)*, rated at not more than 30 mA to the mains system (it may be fitted inside the outlet box or multiple socket). RCDs more sensitive than 30 mA need to be used only if the leakage current is expected to remain below 30 mA, which is rarely the case.

* Sometimes called residual current breaker, RCB, or residual circuit current breaker, RCCB.

1.1.12 Soldering

Soldering serves two functions: it provides (a) the requisite electrical connections, and (b)

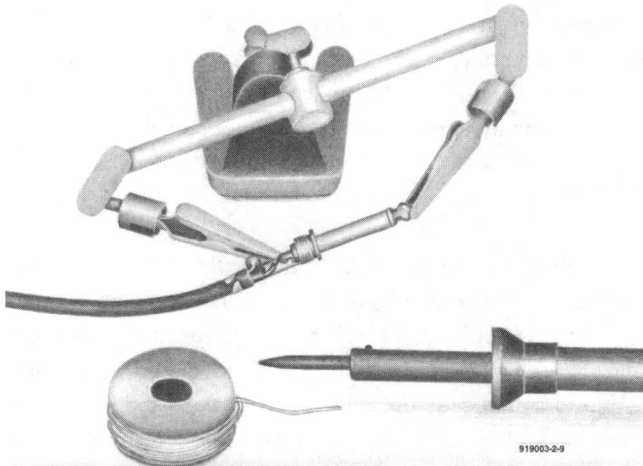

919003-2-9

Figure 1.1.34. Soldering equipment for the sound technician: small (30 W) soldering iron with fine tip, solder tin, and a small retaining tool (here firmly 'holding' a 6.3 mm audio connector).

mechanical support, holding the components of an assembly together.

Making a good soldered joint requires that the parts to be soldered are clean, free of insulation such as enamel, and positioned to remain relatively immobile. The surfaces of the parts must then be heated with a soldering iron, whereupon a good quality solder (an alloy of tin and lead) is to be applied to the heated surfaces, after which the solder must be allowed to cool and solidify. It is important to use solder very sparingly.

It should be noted that solder only adheres to certain surfaces, mainly metallic, and does not adhere to insulating surfaces.

1.2 Quantities, units and symbols

Judging the quality and certain other aspects of sound equipment is facilitated by a number of dimensions, units and definitions.

Some important definitions are listed below.

Sound intensity, I, is the energy in watts passing through an area of one square metre perpendicular to the direction of propagation. The instantaneous sound intensity, I, is

$$I = pu = p^2/\rho c,$$

where p is the instantaneous sound pressure, u is the particle velocity, ρ is the fluctuating density, and c is the speed of sound. Note that the quantity ρc is the characteristic impedance of the medium; in air, $\rho c = 415$ N m^{-2} s at 20 °C.

Sound (or acoustic) pressure, p, is measured in Pascal (Pa), which is one newton per square metre (1N m^{-2} = 10 μbar = 10 dyne cm^{-2}).

Sound power, P_m, of an idealized sound source (monopole) is

$$P_m = \rho c k^2 Q^2/8\pi,$$

where Q, the strength of a monopole source, $= 4\pi a^2 U$, (in which a is the radius of the monopole and U the simple harmonic velocity), $k =$ the wavenumber $= \omega/c = 2\pi/\lambda$. (The expression $\omega/c = 2\pi f/c = 2\pi/\lambda$, or $\lambda = c/f$ relates the speed of sound to the wavelength and the frequency of the sound—this will be reverted to later in this chapter).

Since the range of sound pressure and sound power experience in practice is very large, logarithmic rather than ratiometric (or linear) measures are often used. The most common of these is the decibel. Each quantity in decibels is expressed as a ratio relative to a reference sound pressure, power, or intensity. Whenever a quantity is expressed in decibels, the result is know as a level.

Sound intensity level, L_i, is the sound intensity expressed in decibels compared to a reference of 10^{-12} W m^{-2} (0 dB), which is the faintest sound at 1 kHz that can be heard by a young person. This is expressed as

$$L_i = 10\log_{10}(I/I_r)$$

where L_i is the intensity level in dB, I = intensity in watts, $I_r = 10^{-12}$ watts.

Sound pressure level, L_p, (sometimes referred to as SPL) is the sound pressure expressed in decibels relative to a reference pressure, p_r, which is taken as 20 μPa or 2×10^{-5} N m^{-2}. In formula form:

$$L_p = 20\log_{10}(p/p_r)$$

where p = pressure in μPa.

The sound power level, L_w, is given by

$$L_w = 10\log_{10}(W/W_r),$$

where W is the sound power of a source and $W_r = 10^{-12}$ watts is the reference sound power.

Loudness. The perception of loudness depends on the sound pressure level and frequency. Sounds at the low or high end of the spectrum seem less loud than those in the middle region because of the ear's varying sensitivity at different frequencies.

The phon is a unit of loudness which takes into account the dependence on frequency of the ear's sensation of loudness. One phon is equivalent to an SPL of 1 dB at 1 kHz. When sound pressure levels are quoted at 1 kHz, they are equal to the number of phons.

The transfer function A of any part of the signal path (connecting cable, amplifier, effects unit, and so on) is defined as the ratio of the output quantity to the input quantity; for instance, the ratio of the output voltage, U_2 to the input voltage, U_1, that is,

$$A = U_2{:}U_1.$$

When A1, there is amplification; when A<1, there is attenuation.

Transfer and transfer function are, of course, meaningless unless they are expressed by a quantity. In the case of an amplifier, the transfer is usually called amplification. An index then indicates whether voltage ampliciation, A_v, or current amplification, A_i, is meant.

All passive parts of a signal path (connecting cable, crossover filter, plug-and-socket connections, and so on) cause attenuation of the signal. Active components usually provide amplification (amplifiers, boosters, and so on). In effects units (flanger, phaser, chorus, and so on), care must be taken to ensure that A = 1. This means that the unit can affect the shape and properties of the signal, but not its amplitude. Unfortunately, this is not often the case.

In sound engineering, in the electronic as well as in the acoustic parts, quantities are used whose numerical values differ by powers of ten (or two). For instance, the output voltage of an electric guitar, depending on the type and the manner it is played, varies from a few microvolts (μV; $1\ \mu V = 10^{-6}\ V = 0.000001\ V$) to some hundreds of millivolts (mV; $1\ mV = 10^{-3}\ V = 0.001\ V$). Useful sound levels vary from $10^{-6}\ W$ (normal speech between two people) and $10^8\ W$ (a rocket being launched). Since such quantities are rather large to use in our daily work, it has been found much more convenient to use logarithmic ratios. A difference is made between the transfer ratio (when A1) and the attenuation ratio (when A<1). In general, the transfer ratio is defined as

$$a = 20\log_{10} A \qquad\qquad [Eq.\ 18]$$

The voltage transfer ratio (in amplifiers, the voltage amplification) is calculated as

$$a_u = 20\log_{10}(U_2/U_1) \qquad\qquad [Eq.\ 19]$$

Electrical or acoustical power ratios are expressed by

$$a = 10\log_{10} P_2/P_r \qquad\qquad [Eq.\ 20]$$

Since a is a dimensionless quantity, it was given the pseudo-quantity bel (B) by engineers at the Bell Laboratories in the USA in commemoration of Alexander Graham Bell, the Americanized, Victorian Scotsman who invented the telephone and was the first electro-acoustic engineer. Since the bel is an impractical large unit, a tenth of it, the decibel (dB) is invariably used in practical calculations.

Example: If a microphone preamplifier has a gain of 20 dB, what is the amplification? To solve this, use Eq. 19:

$$antilog_{10}\ a/20 = U_2/U_1 = A$$

$$\therefore A = 10^{a/20} \qquad\qquad [Eq.\ 21]$$

So, the voltage amplification is

$$A = 10^{20/20} = 10.$$

When A<1, the transfer function becomes negative, and this means that the signal is

dB	Voltage ratios		Power ratios	
	U_2/U_1	U_1/U_2	P_2/P_1	P_1/P_2
100	10^5	10^{-5}	10^{10}	10^{-10}
80	10^4	10^{-4}	10^8	10^{-8}
60	10^3	10^{-3}	10^6	10^{-6}
40	100.0	0.01	10.0	0.0001
30	31.623	0.03162	1000.0	0.001
20	10.0	0.1	100.0	0.01
10	3.1623	0.31623	10.0	0.1
6	1.9953	0.5012	3.9811	0.2512
3	1.4125	0.7079	1.9953	0.5012
2	1.2589	0.7943	1.5849	0.631
1	1.22	0.8912	1.2589	0.7943

Note the following decibel terms.

dBA – A-weighted sound pressure level.

dBC – C-weighted sound pressure level.

dBm – Power ratio relative to 1 mW.

dBq – Absolute voltage level of noise in an audio channel relative to a reference voltage defined by international standard.

dBr – dB ratio relative to some specified reference level.

dB_{SPL} – Sound Pressure Level, referred to nominal threshold of human hearing (0 dB).

dBu – dB ratio referred to a level of 0.775 V, any impedance.

dBv – dB ratio relative to a level of 1 volt.

dBw – dB ratio relative to a level of 1 watt.

dBµ – dB ratio relative to a level of 1 µV

Table 1.2.1

attenuated. The attenuation ratio is also expressed in dB and is calculated by reversing the numerator and denominator, that is, the input becomes the numerator and the output the denominator ($A = U_1/U_2$).

These relationships are of great importance in electro-acoustic engineering. It is convenient to work with ratios greater than unity. Should $U_2 < U_1$ or $J_2 < J_1$, invert the ratio and precede the logarithm with a negative sign (indicating attenuation or loss).

There are other other decibel scales—see Table 1.2.1. One is the acoustic, in which the zero level is taken at $I_o = 2 \times 10^{-5}$ N m^{-2} = 20 μPa (sound pressure scale) or $I_o = 10^{-12}$ W m^{-2} (sound intensity scale). The other is the absolute electrical (dBm) scale, in which the zero level is $P_o = 1$ mW; any power P_x has a dBm level of $10\log_{10}(P_x/10^{-3})$. A standard communication impedance of 600 Ω is assigned to this scale giving a corresponding voltage scale. The zero level when 1 mW is dissipated in 600 Ω is $U_o = 775$ mV. The dB level of any voltage U_x on this scale is $20\log_{10}(U_x/775 \times 10^{-3})$. For convenience, a number of voltage and power ratios and the corresponding decible values are given in Table 1.2.1. Intermediate values may be derived from the characteristics in Figures 1.2.1. and 1.2.22.

An important advantage of using decibels is that transfer ratios of the various part in a signal path can simply be added to give the transfer ratio of the entire path.

Example: Figure 1.2.3 shows a signal path consisting of two cables and a microphone preamplifier. The cables are passive components and can therefore introduce attenuation

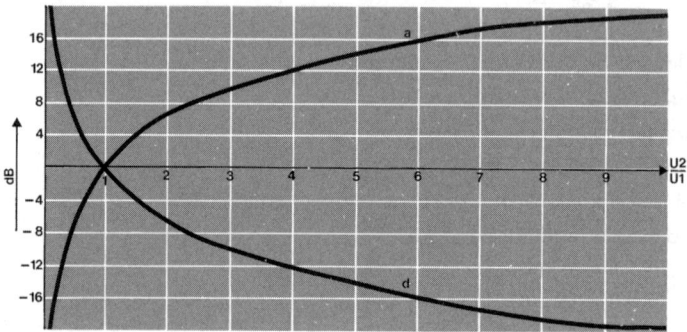

919003-2-10

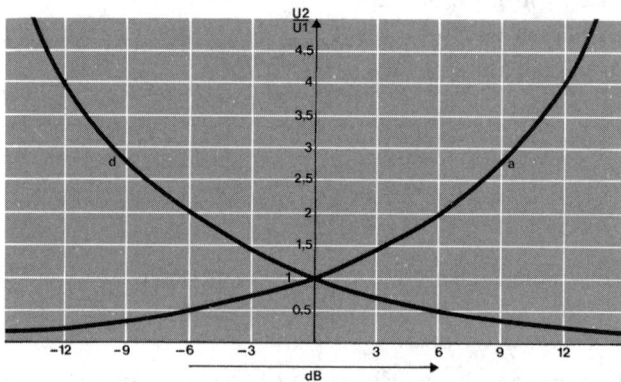

919003-2-11

Figure 1.2.1. Characteristic curves showing the relationship between voltage and sound pressure ratios and the associated decibel values.

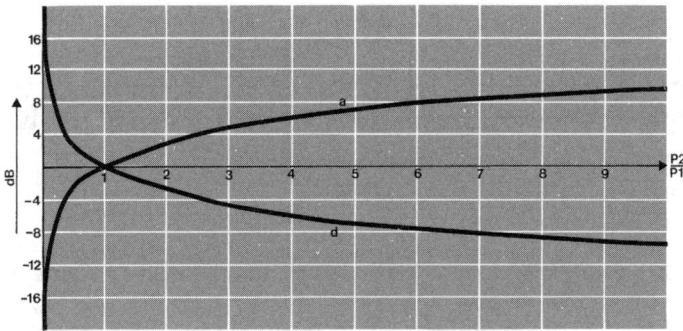

919003-2-12

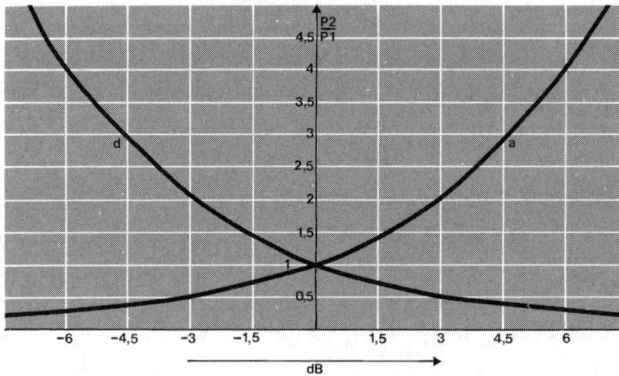

919003-2-13

Figure 1.2.2. Characteristic curves showing the relationship between power ratios and the associated decibel values. It does not matter whether the power is electrical or acoustical.

only. The total transfer ratio is calculated by adding the transfer ratios of the various components:

$$a_t = a_1 + a_2 + a_3 = (-2) + (+24) + (-2) = 20 \text{ dB}$$

This shows that the overal transfer is an amplification of the microphone signal. From Table 1.2.1, the amplification, $A = \times 10$. If therefore the microphone output is 1 mV, the output voltage, $U_o = 10$ mV.

Example: The sound pressure level J of a loudspeaker is, according to the manufacturer's data sheet, 90 dBμ at 1 metre from the diaphragm. What is the sound intensity?

$$I = I_o \times 10^{90/20} = 20 \times 10^{-6} \times 10^{90/20} = 0.63 \ \mu\text{Pa}.$$

39

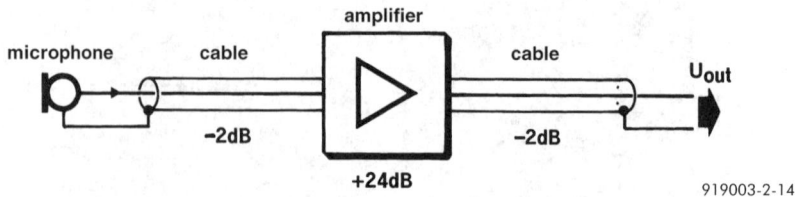

Figure 1.2.3. Example of a reproduction link showing associated transfer ratios.

Note that raising the sound pressure by 6 dB doubles the sound intensity.

Logarithmic quantities such as the decibel are often used to good effect in graphical characteristics. This makes it possible for quantities that differ by a factor 10^3 or 10^4 to be included in the characteristic without this becoming impractically long. This factor should always be borne in mind when a graph is being studied, because it makes frequency response characteristics, for instance, look much flatter than they really are.

As an example, Figure 1.2.4 shows the frequency vs sound pressure level characteristic of a small loudspeaker box with a volume of only 4 litres. The characteristic is based on measurements taken in a anechoic chamber at a distance of one metre from the loudspeaker. The scale on the y-axis is in 5 dB steps. The level of a sound at 20 Hz is 2.5 dB, while that of a sound at 500 Hz is 14 dB. This means that the sound at 500 Hz is reproduced 11.5 dB stronger than that at 20 Hz. With Eq. 21, the difference in amplification can be calculated:

$$A = 10^{11.5/20} = 3.75.$$

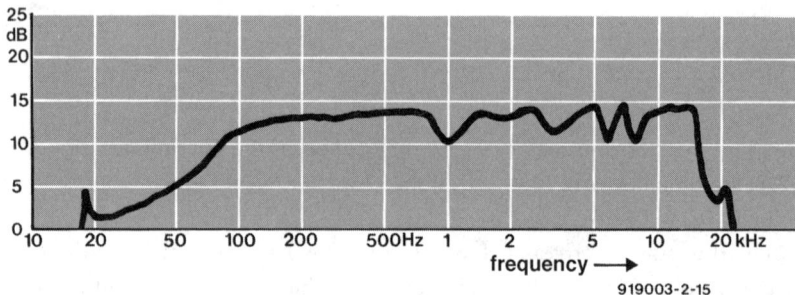

Figure 1.2.4. Characteristic curve showing the relationship between sound pressure and frequency of a small loudspeaker.

So, the amplification at 500 Hz ɪs 3.75 times greater than that at 20 Hz.

Frequency response characteristics give detailed information on the transfer function of an equipment. Yet, or therefore, some manufacturers only divulge the lower and upper frequencies that can be processed by an equipment. For instance, 'frequency range from 60 Hz to 16,000 Hz ±2 dB' means that the transfer function in this range can vary from –2 dB to +2 dB from a flat response. When a frequency range is given without indication of the maximum deviation from the ideal response, this traditionally means that there may be a variation of up to –3 dB.

Apart from the usual Cartesian coordinates (that is, with two axes at right angles), polar coordinates are also often used in electro-acoustic engineering. The characteristic is then given as a function of the direction or angle. More will be said about this in the chapter on microphones.

1.3 Fundamentals of acoustics

The ultimate aim of each and every musical activity is the production of audible vibrations of the air. In general, mechanical vibrations of an elastic medium, such as air, are called sound; when the fluctuations lie in the audio band, they are called audible sound. The study of these vibrations of air is called acoustics. The sound heard by an audience during a performance in a hall depends in the first instance on how well acquainted the sound engineer is with acoustics. The sound installation is of lesser importance. Simple instruments and voice amplifiers can, when the acoustics of the hall is taken into account, give an equally well defined sound as can be obtained with an extensive installation. This may sound odd to some readers, but it can be explained by the principles of sound engineering and transfer technology. For better or for worse, the concert hall cannot be modified as far as acoustic properties are concerned, and the same applies to the available amplifier equipment. The art lies in making optimum use of the acoustic properties of the hall and the sound installation.

1.3.1 Vibrations and waves

From a purely scientific point of view, sound is nothing but a wave motion, that is, a periodically changing of pressure in an elastic medium. This medium may be solid, liquid or gaseous. The distances between the particles of the medium (atoms or molecules) become periodically greater and smaller. The particle itself vibrates around a more or less fixed position. Transport of parts or material does not take place. Only the sound energy is transported in the direction of propagation. This is clarified in Figures 1.3.1 and 1.3.2.

Expanding sound waves need a medium to travel in. In a vacuum (as in space, for

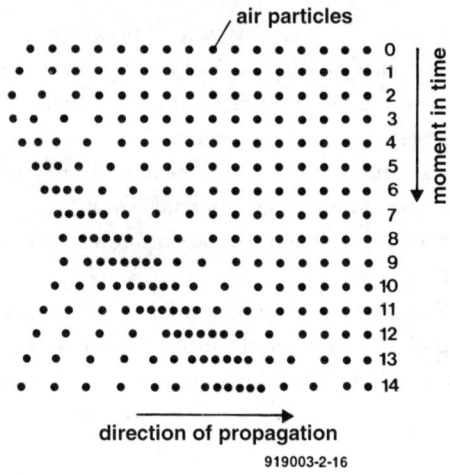

air particles

moment in time

direction of propagation

919003-2-16

Figure 1.3.1. This is how we may imagine that air particles are set into vibration by a sound wave. A particle that vibrates around its rest position excites adjacent particles into vibration until all particles are vibrating.

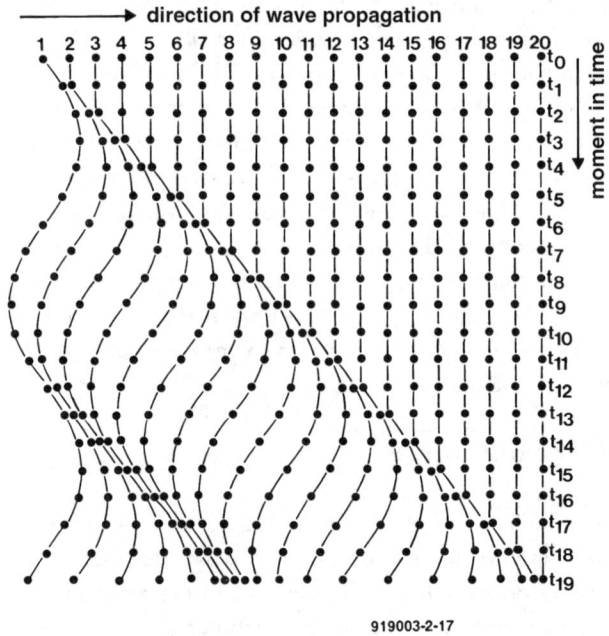

direction of wave propagation

moment in time

919003-2-17

Figure 1.3.2. Same particles as in Figure 1.3.1 but here interconnected by lines to clarify the wave motion.

42

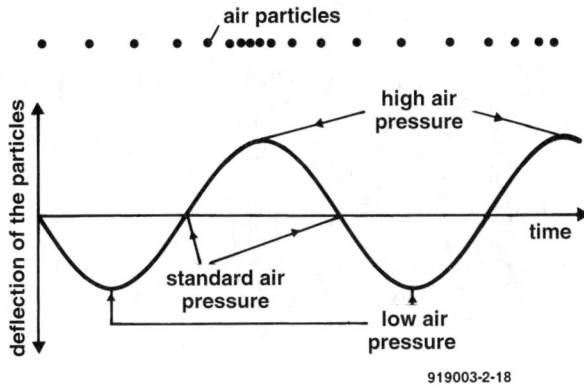

Figure 1.3.3. Within a given volume of air at standard air pressure,
a sound wave causes positions of high and low air pressure.

instance), there cannot be sound because there is no medium via which the waves can move (contrary to what some science fiction books and films want us to believe). When a single particle which is made to vibrate by sound energy is observed, it appears that it vibrates around a rest position. The vibration may be described by two variables: deflection (amplitude) and time. In the simplest case, the vibration is a sine wave—see Figure 1.3.3. The quantities amplitude, frequency and period already discussed in 1.1.3 are indispensable here, too.

In electro-acoustics it is also important to know the wavelength, λ, of a sound vibration. As stated earlier, this is given by:

$$\lambda = c/f,$$

where c is the speed of sound in metres per second, m s^{-1}, f the frequency in hertz (Hz), and λ the wavelength in metres, m. Note that the speed of sound, unlike that of light, is not a constant. It depends on atmospheric pressure and, more particularly, temperature. At 0 °C, it is 331.6 m s^{-1} (1090 ft), and rises by 60 cm (about 2 ft) for each degree C increase in temperature. Thus, at room temperature (20 °C or 68 °F), it is about 344 m s^{-1} (around 1130 ft). More about this later.

The wavelengths at which sound becomes audible (to humans) lie between 0.02 m (highest frequency) and 20 m (lowest frequency). This is clarified by Figure 1.3.4, which shows the wavelength of a sinusoidal vibration in air at a frequency of 1000 Hz.

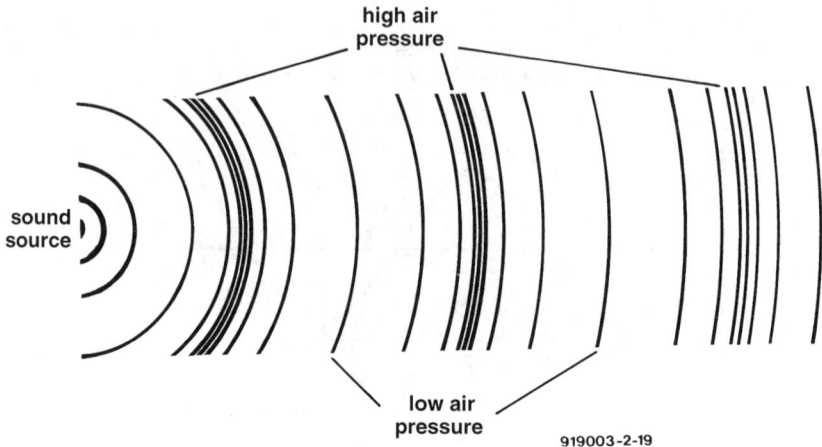

high air pressure

sound source

low air pressure

919003-2-19

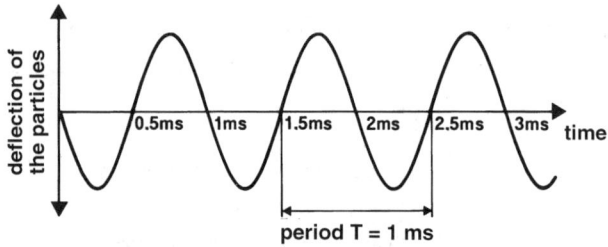

deflection of the particles

0.5ms 1ms 1.5ms 2ms 2.5ms 3ms time

period T = 1 ms

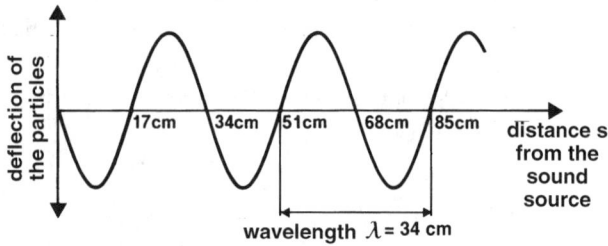

deflection of the particles

17cm 34cm 51cm 68cm 85cm distance s from the sound source

wavelength λ = 34 cm

919003-2-20

Figure 1.3.4. The upper illustration is a diagrammatic representation of areas with high and low pressure. The graphs show the movement of the air particles (that is, air pressure) as a function of time and as a function of the distance between the areas of high and low pressure and the sound source. The period, T, and the wavelength, λ, may be determined from the distance between the tops of the graphs

44

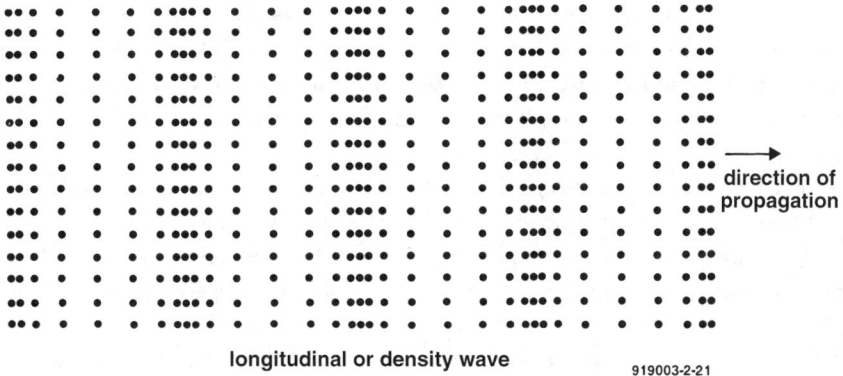

longitudinal or density wave

919003-2-21

Figure 1.3.5. Behaviour of air particles in a longitudinal wave.
The particles move in the same direction as the wave. The distance
between individual particles changes continuously.

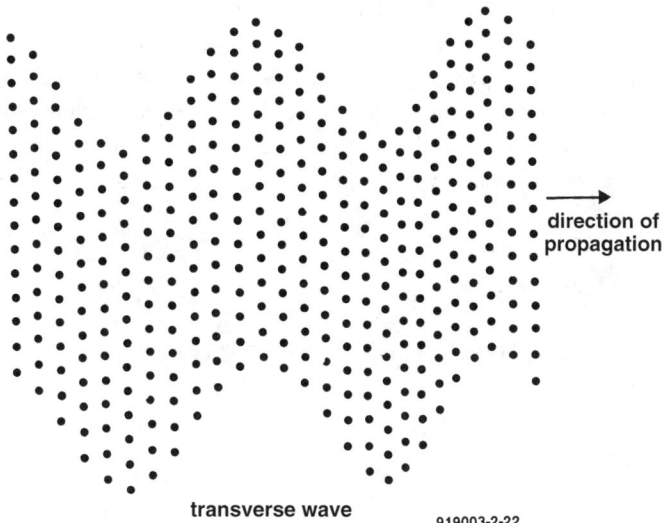

transverse wave

919003-2-22

Figure 1.3.6. Behaviour of air particles in a transverse wave (produced, for instance,
by a vibrating string). The distance between discrete particles remains constant.
Instead of air rarefaction and compression, spatial distortion occurs.

The sum total of vibrating particles in a room is not called vibrations, but waves. Waves cannot be described properly by amplitude and time alone; their kind is also an important aspect. In gases and lliquids, sound is propagated by longitudinal (densitiy) waves. The direction of propagation of individual particles is the same as that of the wave itself—see Figure 1.3.5.

In solids, the direction of propagation of many individual particles is at right angles to that of the wave. This kind of wave is called transverse—see Figure 1.3.6.

With few exceptions, such as sound boards and stretched strings (violins, cellos), this book is concerned with sound propagation in air only, and thus with longitudinal waves. Moreover, it deals only with audible sound, that is, frequencies from about 20 Hz to 20 kHz. Sound below 20 Hz is called infrasound, and that above 20 kHz, ultrasound.

1.3.2 Sound field

When air is set into vibration by a monopole (point sound source), the sound pressure areas move away from the source like expanding spheres, so creating a sound field. The energy is distributed evenly over the surface of each sphere and so is spread over an increasing area. Monopoles do not occur in practice: sound sources such as instruments and loudspeakers radiate the sound at different intensities, depending on frequency, in various directions. Nevertheless, the idea of a monopole is very convenient to make clear in a not too complex way how sound is propagated. The ideal spherical expansion of sound pressure is approached at very low audible frequencies. At these frequencies, theoretical considerations may often be applied in practice without detriment.

It has already been stated that the speed of sound is not a constant, but depends on the atmospheric pressure and temperature. In most cases, the atmospheric pressure is ignored, and the speed is then expressed by

$$c = 331.6(1 + t/273)^{1/2} \qquad \text{[Eq. 25a]}$$

where c is the speed of sound in m s^{-1} at 0 °C, and t is the ambient temperature in °C, or, more simply

$$c = 331.6 + 0.6 t \qquad \text{[Eq. 25b]}$$

Example. Since the wavelength of the sound waves emanating from a wind instrument is determined by its geometry, its frequency is given by Eq. 24 and Eq. 25:

$$f = c/\lambda = 331.6/\lambda + 0.6 t/\lambda.$$

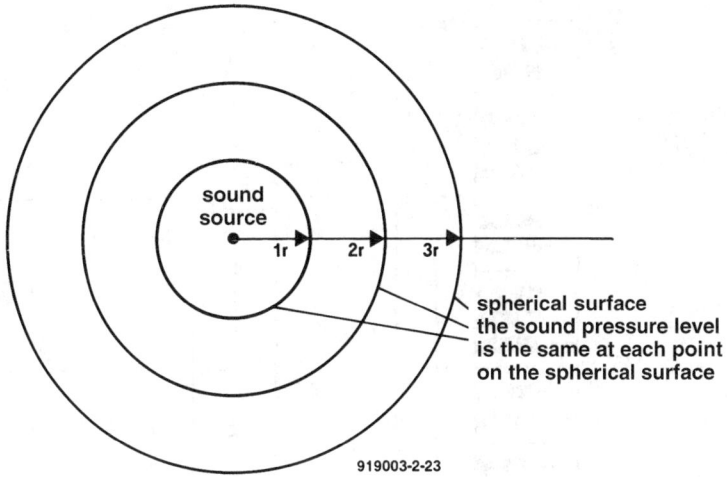

919003-2-23

Figure 1.3.7. When a sound source radiates uniformly in all directions,
sound pressure areas emanate from it like expanding spheres. The energy
is distributed evenly over the surface of the sphere.

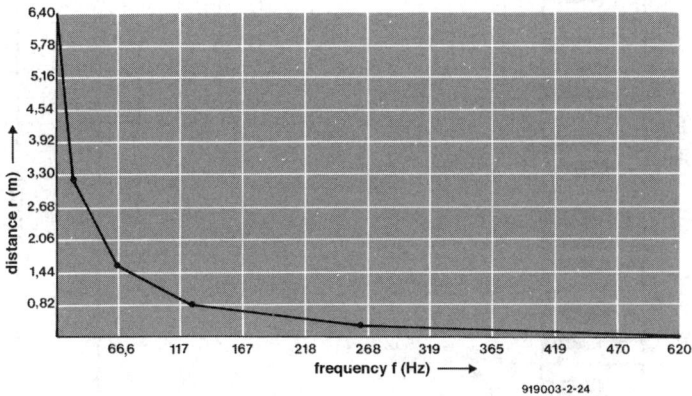

919003-2-24

Figure 1.3.8. The higher the frequency, the earlier the sound pressure level increases.
If this is not allowed to exceed 3 dB (that is, about ×1.4),
the minimum distance as shown here should be observed.

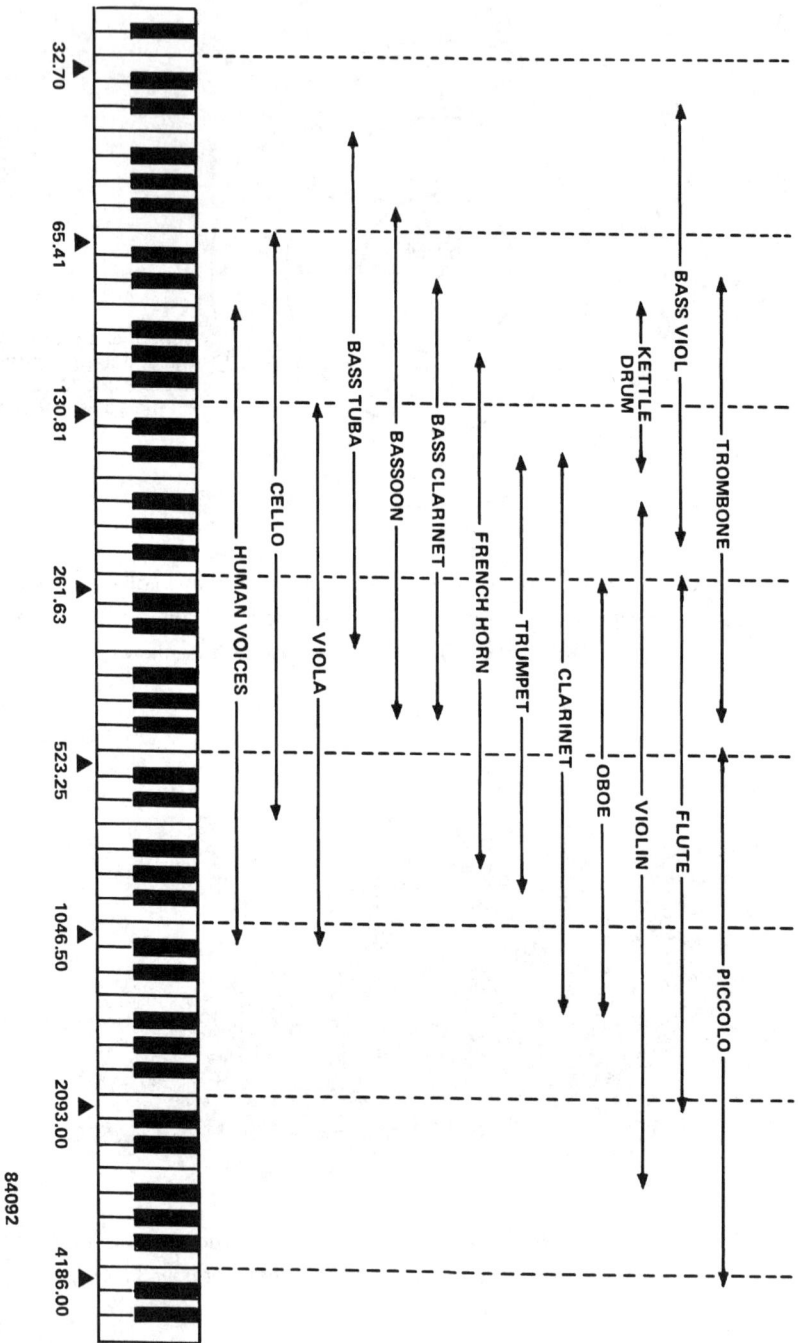

Various musical instruments and their frequency ranges in hertz over eight octaves of the chromatic scale.

84092

48

tone octaves	C	C#	D	D#	E	F	F#	G	G#	A	A#	B
+4	4186.00	4434.91	4698.62	4978.02	5274.05	5587.64	5919.90	6271.91	6644.86	7039.99	7458.60	7902.12
+3	2093.00	2217.46	2239.31	2489.01	2637.00	2793.80	2959.93	3136.00	3322.48	3520.00	3729.31	3951.10
+2	1046.50	1108.73	1174.70	1244.55	1318.50	1396.90	1479.96	1568.00	1661.24	1760.00	1864.66	1975.50
+1	523.25	554.36	587.33	622.25	659.26	698.46	740.00	784.00	830.61	880.00	932.33	987.77
0	261.63	277.19	293.66	311.12	329.63	349.23	370.00	392.00	415.31	440.00	466.16	493.88
−1	130.81	138.59	146.83	155.56	164.81	174.61	185.00	196.00	207.65	220.00	233.08	246.94
−2	65.41	69.30	73.42	77.79	82.41	87.31	92.50	98.00	103.83	110.00	116.54	123.47
−3	32.70	34.64	36.71	38.89	41.20	43.65	46.25	49.00	51.91	55.00	58.27	61.74
−4	16.35	17.32	18.35	19.45	20.60	21.83	23.12	24.50	25.95	27.50	29.13	30.87

Table 1.3.1. Tones and their equivalent frequencies in hertz over eight octaves of the chromatic scale.

The frequency, or pitch, increases in direct proportion to the speed of sound and thus with ambient temperature. When the ambient temperature rises by 10 per cent, the pitch increases by just under two per cent. Since normal room temperature is 20 °C, a pitch at 10 °C sounds too low by about 2 per cent and at 30 °C, about 2 per cent too high. With the usual division of an octave in twelve semitones (chromatic scale—see Table), the change in frequency between two semitones is six per cent. When the temperature rises or drops by 10 per cent, a string instrument is detuned by about 1/3 of a semitone. This is why string instruments have to be tuned with the aid of a wind instrument or a tuning fork. Note that 'electronic tuning forks' are in general not accurate enough.

In an infinitely large space, the sound field consists of a near field in the immediate surrounding of the sound, and a far field, which is what we normally listen to. When the source is radiating in an enclosed area (room or hall), reflections from the boundary walls and ceiling create a reverberant field that is superimposed on the near and far fields. In an ideal case, the sound energy in such a composite field is distributed evenly throughout the enclosed area: there is no preferential direction of propagation. When the contribution of the reverberant field increases, it becomes more and more difficult to pinpoint the location of the source: in a pure reverberant field, it is quite impossible. A practical sound field is always a composite one. The reverberant field is particularly noticeable in large halls, churches and cathedrals. However, to obtain a transparent sound picture, the reverberant contribution must not exceed a certain value. A pure direct field can be obtained only in an anechoic chamber. The sound of such a field is dry and does not present a coherent picture.

Another important quantity is the characteristic acoustic impedance, ρc, which depends entirely on the dimensions and nature of the medium. At standard atmospheric pressure and an ambient temperature of 20 °C its value is about 415 N s.

As mentioned earlier, a monopole radiates sound equally in all directions, and the sound wave emanating from it expands like a sphere. In such a wave, the sound pressure and the velocity are not in phase, which makes calculations of the sound field more intricate.

When the radius of a sphere is very long, the surface of the sphere becomes flat for all practical purposes. The length of radius where the sphere may be considered flat is much shorter at high frequencies than at low ones. It appears that the phase shift between sound pressure and velocity is particularly pronounced near the source. In certain circumstances, the frequency may be much higher close to the source. This is a factor (the proximity effect) which will be reverted to in Chapter 2. The increase in velocity at a given point depends on the frequency (pitch).

The relationship between frequency, f, and distance, r, from the source is shown in Figure 1.3.8. Note that owing to the proximity effect, the pressure level rises by about 3 dB. In the following calculations, a plane wave is assumed. Such approximations do not invalidate the practical side.

Imagine that a sound source radiates its energy (or sound power), W, spherically. The energy is thus spread over an increasing area, A $(=4\pi r^2)$, as the sphere expands outward. The energy therefore diminishes as the distance, r, in metres from the source increases. The sound intensity, I, at any point from the source is given in watts by

$$I = W/A = W/4\pi r^2. \qquad \text{[Eq. 26]}$$

The sound intensity is also given by

$$I = p^2/\rho c. \qquad \text{[Eq. 27]}$$

It is clear from Eq. 26 that the sound intensity is inversely proportional to the square of the radius or the distance, r, from the source. If it is arbitrarily assumed that at a distance r from the source the sound intensity is I, then at a distance 2r, the intensity is only $I/4$, and at a distance 3r, only $I/9$—see Figures 1.3.9 and 1.3.10.

When the sound source is placed in front of a wall that reflects the sound totally, the energy is radiated hemispherically. Assuming that the source radiates the same power, the sound intensity will be twice that in the previous example:

$$I = 2W/4\pi r^2 = W/2\pi r^2.$$

So, assuming that the sound energy is radiated over a hemisphere and that the distance from the source is sufficient to consider the sound wave plane, the sound pressure as a function of the distance r can be calculated from:

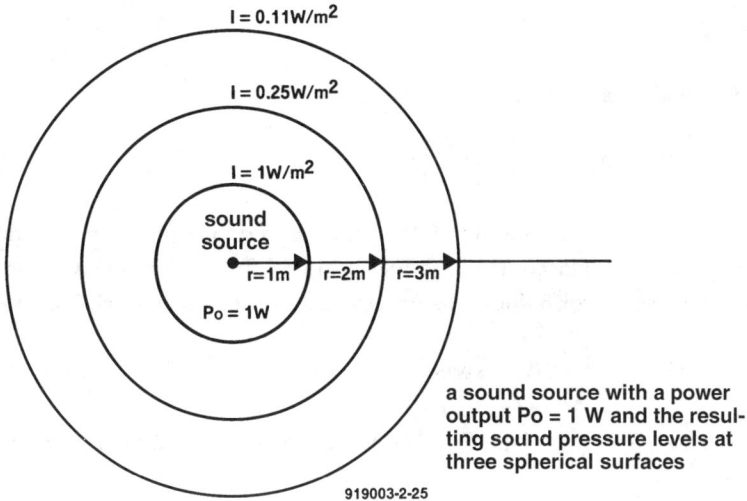

I = 0.11W/m²

I = 0.25W/m²

I = 1W/m²

sound source

r=1m r=2m r=3m

Po = 1W

a sound source with a power output Po = 1 W and the resulting sound pressure levels at three spherical surfaces

919003-2-25

Figure 1.3.9. The sound intensity I is inversely proportional to the square of the distance.

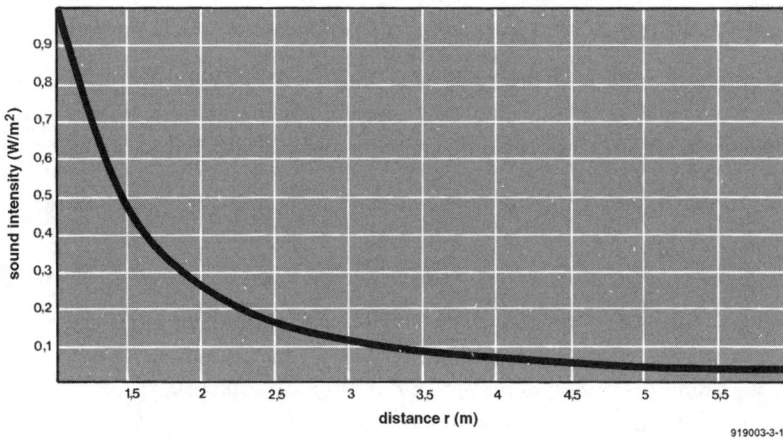

919003-3-1

Figure 1.3.10. Characteristic curve of the sound intensity I as a function of distance. It is based on a sound intensity of 1 W m⁻² at a distance of 1 metre from the source.

51

$$p = (\rho cW/2\pi r^2)^{1/2},$$ [Eq. 28]

or, the sound pressure level is

$$L_p = 20\log_{10}[(\rho cW/2\pi r^2)^{1/2}]/p_o$$ [Eq. 29]

Example. An electrical power of 30 W is applied to a loudspeaker having an efficiency of 10 per cent that is placed against a reflecting wall. The sound energy is radiated forward over a hemi-spherical surface. The acoustic power is calculated with Eq. 11:

$$P_o = \eta P_i = 0.1 \times 30 = 3\ W.$$

Substituting this value in Eq. 29 to find the SPL at 1 metre distance from the loudspeaker gives:

$$L_p = 20\log_{10}[(415 \times 3)/2 \times 3.142 \times 1^2)^{1/2}]/2 \times 10^{-5} \approx 117\ dB.$$

At 10 metres from the loudspeaker, the SPL is

$$L_p = 20\log_{10}[(415 \times 3)/2 \times 3.142 \times 10^2)^{1/2}]/2 \times 10^{-5} \approx 97\ dB.$$

At a distance of 700 metres, the SPL is still 60 dB.

Another example. In a music hall, the minimum SPL at the back row must be 70 dB. The distance between the back row and the stage is 30 m. The sound is reflected totally by the wall behind the stage. What is the minimum acoustic power that must emanate from the stage?

$$L_p = 20\log_{10}p/p_r$$

$$\therefore p = 10^{SPL/20} \times p_r = 10^{70/20} \times 2 \times 10^{-5} = 0.06325\ N\ m^{-2}.$$

This gives the sound pressure that must obtain at 30 m from the stage. Since the sound power is distributed over a hemisphere, from Eq. 28

$$W = 2\pi p^2 r^2/Z_a = 2 \times 3.142 \times 0.06325^2 \times 30^2/408 = 55.4\ mW.$$

Remember, this is acoustic, not electric, power.

Some sound sources and their associated SPLs are given in Table 1.3.2.

Since it is often of value to be able to estimate the order of magnitude of sound

SOUND	L_p (dB)
threshold of hearing	0
rustling of leaves, faint whisper	10
quiet surroundings	20
ticking of clock, normal whisper	40
normal conversation, tearing of paper	50
typewriter, vacuum cleaner, sounds in office	60
sounds in street	70
shouting, busy traffic, motorcycle	80
loud car horn, pneumatic hammer, large orchestra	90
loud noises in a factory, workshop, or printshop	100
siren, jet plane, sand blasting	110
aircraft engine at close quarters	120
threshold of pain	130

Table 1.3.2. A number of sounds and associated sound pressure levels in dB.

SOURCE OF SOUND	P_m(max) in watts
violin	0.001
human voice (average)	0.010
piano	0.200
trumpet	0.300
electric guitar	3.000
percussion instruments	10,000
classical orchestra (75 musicians)	70,000.000
alarm siren	1,000,000.000
rocket being launched	100,000,000.000

Table 1.3.3. A number of sounds and associated power levels in watts. Frequencies
of the various musical instruments are shown in the illustration on page 48.

intensities, here is another example. A loud noise has a sound pressure, p, of, perhaps, 0.1 Pa, which is equal to an SPL of 74 dB. If this SPL is to be obtained at 1 m distance from the source, the sound being radiated forward, the sound intensity required is (Eq. 26 and Eq. 27):

$$W = 2 \times 3.142 \times 0.1^2 \times 1^2/408 = 0.154 \text{ mW.}$$

To obtain the same sound pressure at a distance of 5 m, an intensity of 3.85 mW is needed. To obtain an SPL of 138 dB (which is a grave danger to health) at a distance of

1 m, an intensity of 388.7 W is required, and to obtain this SPL at a distance of 5 m, the sound intensity required is a staggering 9716.7 W. And remember, this is acoustic, not electric, power.

From these examples it is obvious that the dynamic range used in electro-acoustics is large. Table 1.3.3 shows a number of sound sources and their sound intensities.

In practice, it is desirable that the proportion of the direct sound is evenly distributed throughout the concert hall. This means that point sources cannot be used (even if this were possible). What is needed are sources with large radiating surfaces or a large number of smaller sources spread throughout the hall. The best known examples of many sources spread throughout the hall are churches, theatres, and so on. This cannot be achieved at a rock concert. At such concerts and other large-scale performances, use is invariably made of a two-channel public-address installation with two very large loudspeakers. Better results are only possible when the circumstances are more congenial (small to medium size hall). Each instrument then functions as an individual, natural source. When the various instruments are well placed, a broad radiating area comes about almost naturally, resulting in even distribution of sound and, consequently, a clear, transparent sound picture. However, the acoustics of the hall must also be taken into consideration (more about this in Section 1.3.4). PA installations that are too small for the job only make matters worse, since the sound energy is radiated by two (relatively) small loudspeakers. The sound picture then becomes less transparent and there is no longer an even distribution of the sound. This situation will, of course, be even worse in the case of a single-channel PA installation.

1.3.3 Sources of sound

Sources of sound are all those natural and technical sources that impart sound energy to their surrounding medium. Usually, sound is produced by a vibrating membrane, a stretched string or a reed. Since the vibrating part often is not able to produce the sound at full strength, the vibrations are transferred to a resonating body where the sound is shaped and amplified. A typical example is the guitar. In acoustic guitars, the vibrations of the strings are amplified by the soundboard to make the sound clearly audible. An electric guitar requires a special guitar amplifier to assume the functions of the soundboard. Without amplifier, an electric guitar would barely be heard.

A string, reed or membrane normally does not produce a pure tone consisting of a single frequency. Apart from the fundamental frequency, a number of other frequencies are produced, which are called harmonics by the electronics engineer and overtones by the musician. These overtones give the sound its timbre, by which it is possible to tell an oboe from a flute when both are sounding the same note. The soundboard or amplifier adds more overtones and may even amplify or attenuate certain frequencies. How and to

54

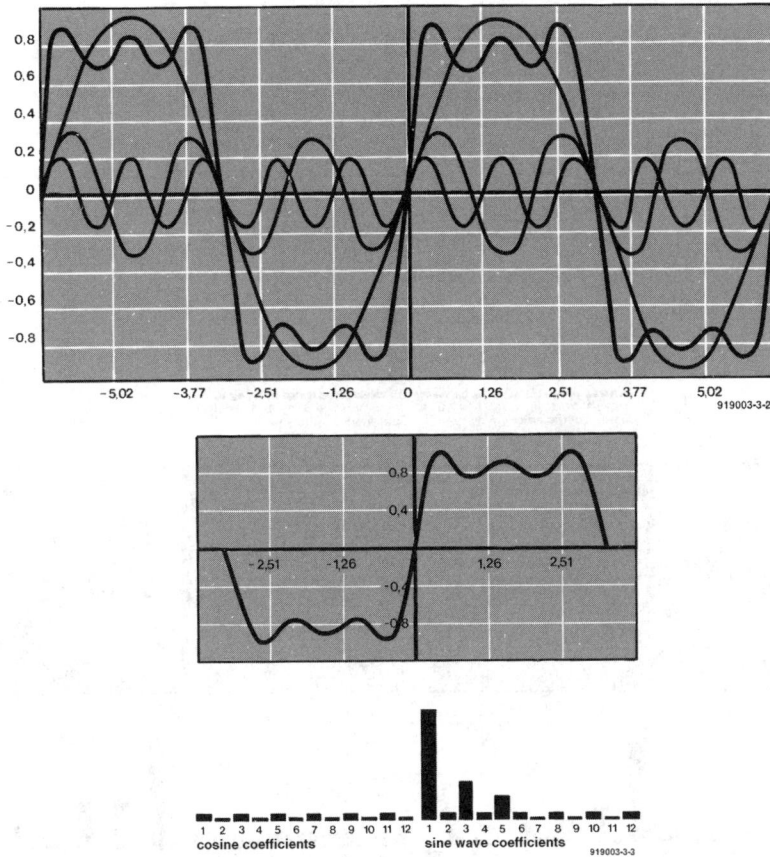

Figure 1.3.11. Synthesis of a square wave from three pure sine waves. The lower bar graphs show the peak value of the discrete harmonics that together form the square wave. The figures below the bar graphs give the frequency ratios of the various harmonics: 1 is the frequency of the fundamental wave. Note that a square wave is formed by the odd harmonics only.

what degree this happens depends on the material and construction of the soundboard or on the amplifier and loudspeakers in case of an electrical instrument. These factors give an instrument its characteristic sound. They are the reason that concert guitars of dissimilar make and model sound differently.

Any waveform, rectangular, sawtooth, triangular, or whatever, consists of a great number of sine waves of different frequency and amplitude. Conversely, a number of sine

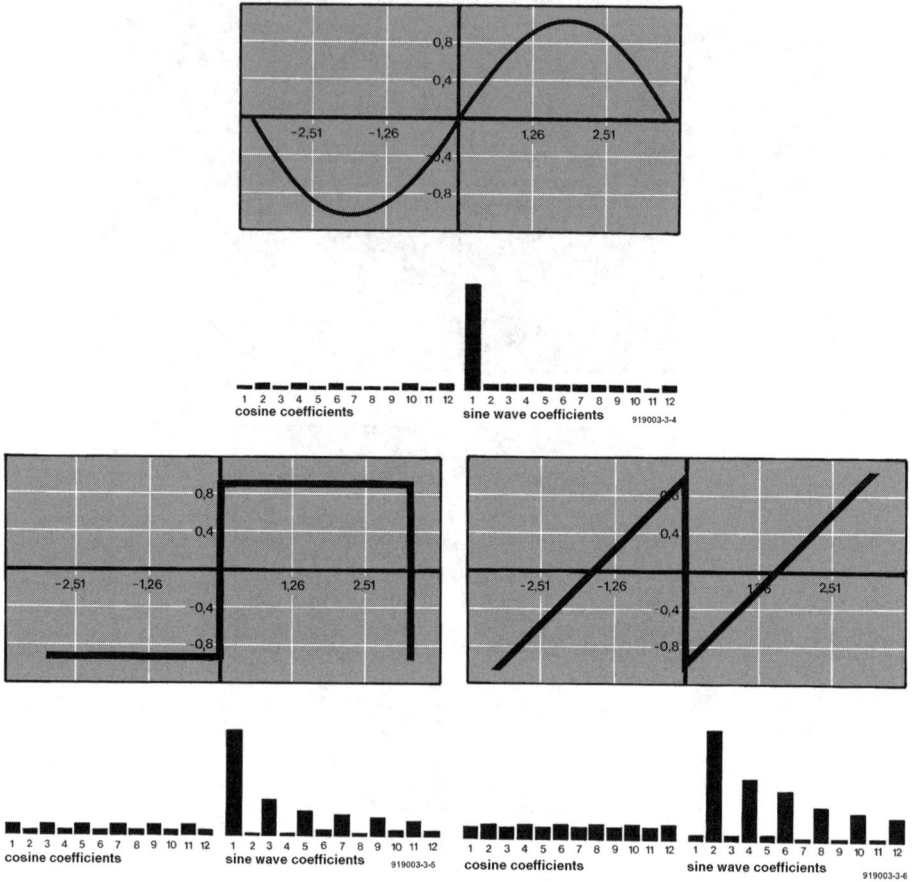

Figure 1.3.12. Analysis of a number of different oscillations. A pure sine wave
has only one component: the fundamental or first harmonic. A square wave
consists of a very large number of odd harmonics. The fundamental is absent.

waves of different frequency and amplitude can be combined to give a complex, non-
sinusoidal waveform. Sounds of one single frequency can be produced only by electronic
instruments. All natural sound sources generate complex sounds. This means that a non-
electronic musical instrument generates a mixture of many sine waves of dissimilar
frequency and amplitude. According to the Fourier theory, any complex waveform can be
analysed into a number of sine waves, called a Fourier series, of different frequency and
amplitude. Figure 1.3.11 shows the first three sine waves of such a series that together

form a square wave, whose shape is already discernible. In practice, it is often useful to be able to estimate the sine waves constituting a certain waveform. Figure 1.3.12 shows the constituents, that is, the frequency spectrum, of a pure sine wave, a square wave, and a sawtooth wave (whose leading edge is called a ramp). It will be noted that a pure sine wave consists of only one frequency: the fundamental. A square wave consists of a fundamental frequency on to which are superimposed a large number of odd harmonics. A sawtooth waveform is composed of the fundamental frequency on to which are

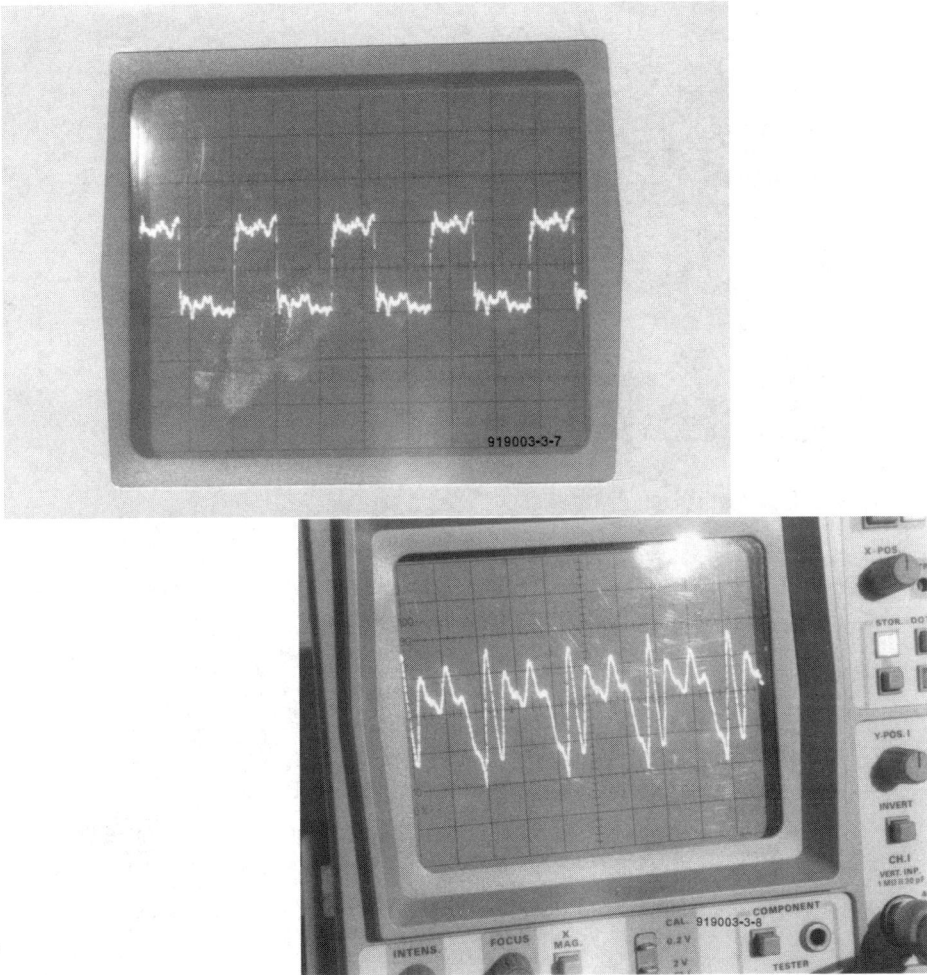

Figure 1.3.13. Upper oscillogram: 440 Hz output signal of a synthesizer (Roland D5, set to solo sound). Lower oscillogram: 440 Hz output of a piano recorded by microphone.

57

superimposed a large number of even harmoncis. Note that the even harmonics have a slightly larger amplitude than the odd ones. As a rule of thumb, waveforms with sharp angles and peaks contain more high frequencies than those that are more 'rounded'.

Square waves, such as produced by certain effects units, contain a very large number of harmonics (in theory, an infinite number). A square wave with slightly rounded corners, as is the case with the output of an overdriven valve amplifier, contains far fewer harmonics, so that the sound is determined to a greater degree by the instrument, the amplifier and the performer. The more overtones are contained, the less the contribution

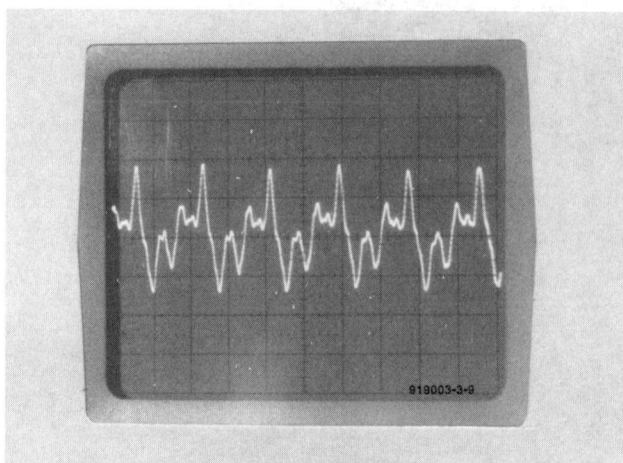

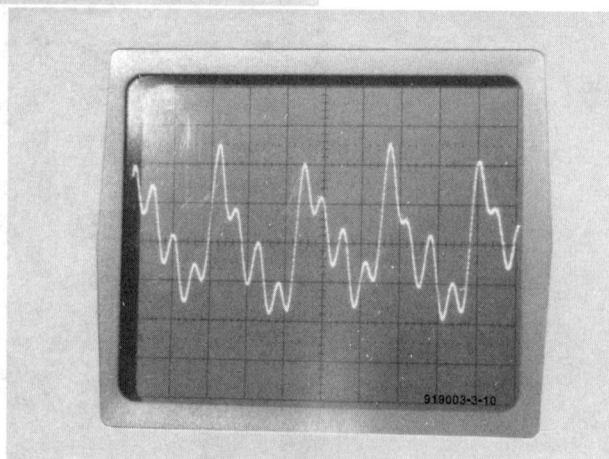

Figure 1.3.14. Top: 440 Hz output signal of an electric guitar;
below: human voice at 440 Hz.

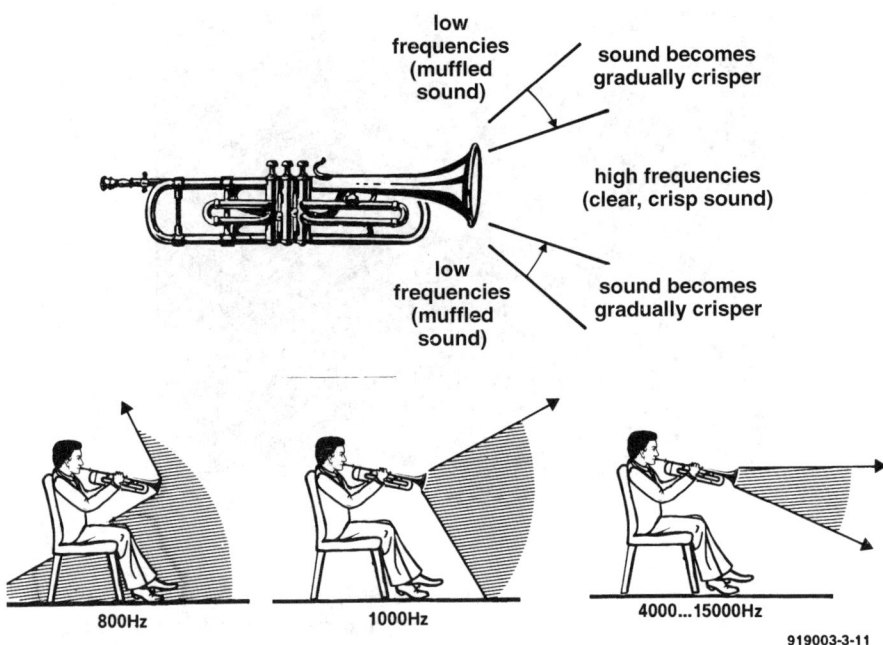

Figure 1.3.15. The radiation pattern of musical instruments is highly dependent on frequency. This example of the trumpet shows that low frequencies are radiated almost spherically, whereas higher frequencies become highly directional.

of the fundamental frequency. In cases of very bad distortion, each instrument and each amplifier sound roughly the same, since the fundamental produced by the instrument is 'drowned' by the overtones. Figures 1.3.13 and 1.3.14 give an impression of the waveforms, and the contribution of the overtones, produced by different instruments. The fundamental frequency in these figure is $f_A = 440$ Hz (international concert pitch). The output of the Roland synthesizer is a square wave and so contains a large number of overtones, whereas that of the piano, recorded via a microphone, shows a large number of peaks and troughs, which indicates a large number of overtones also. In the case of the piano output, the contribution of the overtones , and thus the shape of the signal, alters with the sound level (for instance, when the sound dies away). The same is true of the guitar signal in Figure 1.3.14. Note that the signal produced by a human voice is more rounded: it contains fewer overtones than the output of the instruments.

As mentioned earlier, different sound sources have dissimilar radiation patterns that depend on the wavelength, that is, the frequency. Also, each type of instrument produces a number of harmonics that are unique to it with different amplitudes. From this, it may be

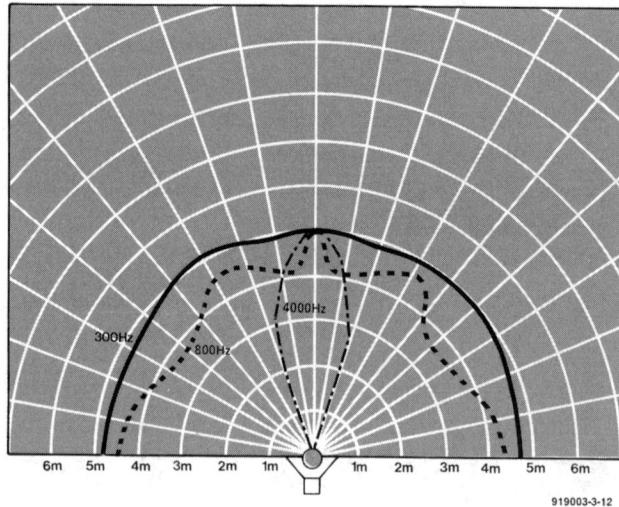

919003-3-12

Figure 1.3.16. The polar diagram of a loudspeaker shows that this is also highly frequency-dependent. Low frequencies are radiated virtually spherically, whereas higher ones become more and more directional.

concluded that each instrument, depending on the tone produced by it, has a different radiation pattern in different directions. In general, this is true for all frequencies above 250 Hz. The higher the frequency, the more the sound is bunched. Low frequencies are generally radiated spherically. This is why the arrangement of the various instruments during a live performance is not critical as far as the bass section is concerned. Bunching of the sound radiation affects timbre, colouration and sound level in the far sound field. This explains why a recording made via microphones often sounds quite different from what has been heard during the recording session. This is particularly true of recordings made with one microphone near the sound source, since only one small section of the sound field has been recorded, while the spatial sound of the instrument has been lost completely. Even when the microphone is moved only one centimetre, the sound may be different. When the sound of an instrument or instruments being recorded is intended to be reproduced by a PA installation, the arrangement of the microphones to ensure a clear, forceful sound is of paramount importance. If, for instance, a microphone is placed directly in front of the diaphragm of a loudspeaker, all high frequencies are reproduced unnaturally strongly, which gives rise to a sharp sound spectrum. Normally, a much better result is obtained when the microphone is placed near the edge of the diaphragm.

An illustrative example of the radiation pattern of a trumpet and of a loudspeaker is given in Figures 1.3.15 and 1.3.16 respectively. It is educational to listen to, and judge for

60

oneself, the impression the sound of an instrument makes at different directions and at various distances.

1.3.4 Room acoustics

As has been mentioned before, the spherical space immediately around a sound source is the near field, and that outside the sphere, the far field. In the near field, particle velocity of the air molecules is often not in the direction of propagation, but in the far field it mainly is. Consequently, propagation in the far field is rather more stable. If the source is radiating in an enclosed room or hall, reflection, refraction, diffraction, scattering and even bunching caused by the boundary walls create a reverberant field superimposed on

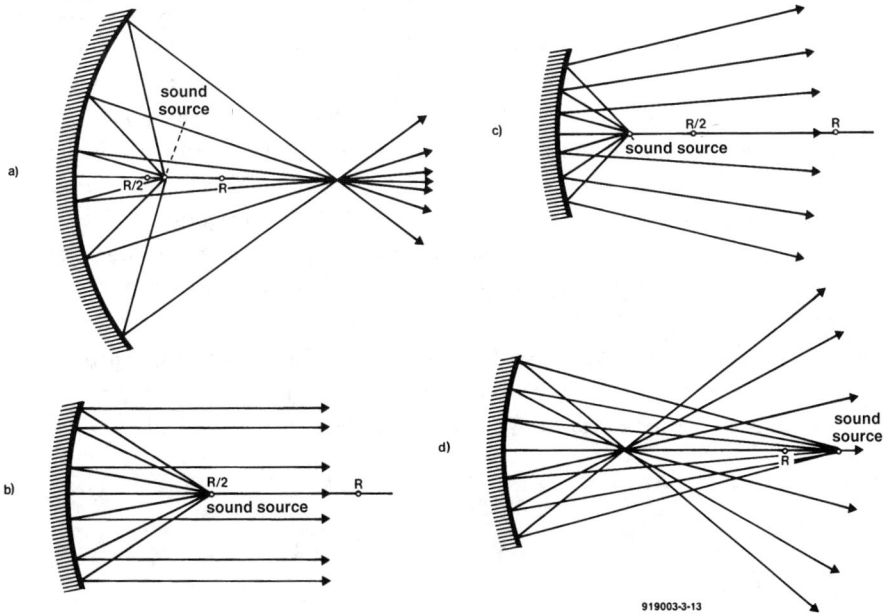

919003-3-13

Figure 1.3.17. When a sound wave is reflected by a concave surface, several effects may be observed. These effects depend on the radius of curvature and the position of the source of sound. In (a) the source is at a distance somewhere between the radius and the half-radius of curvature. The waves are first bunched and then randomly radiated. In (b) the source is at a distance equal to half the radius of curvature: the waves are then reflected in parallel. In (c) the source is at a distance smaller than half the radius of curvature: the waves are radiated divergently. If the source is at a distance greater than the radius of curvature (d), a situation similar to that in (a) arises.

the far field.

When a sound wave is reflected by a plane surface, the exit angle is equal to the angle of incidence; there is no scattering or bunching. If the sound wave is incident at a concave surface, bunching and/or scattering may occur, depending on the wavelength of the sound and the location of the source. Figure 1.3.17 shows various results of a sound wave being reflected by a concave boundary. Convex boundaries always scatter the sound. If the sound wave hits an obstacle that is large compared with the wavelength, the sound wave is reflected (in front) and diffracted (behind) the obstace (rather than scattered).

To make things more complex, the extent of bunching and scattering is generally dependent on the frequency. It is, therefore, not surprising that the reflected, bunched or scattered sound reaches a certain location in the room after the direct sound. It therefore takes a finite time, called the swell time, before the full sound level is observed. The time taken for reflected sound to die away to 1/1000 of the original level of the sound is reverberation time. This is an important aspect of the 'acoustics' of a room. It is responsible for giving body and volume to the sound, for, without it, the sound is weak and thin, as it would be in open air. However, an excess of reverberation would cause speech to sound slurred and music, muddled. The way sound pressure builds up at a certain location in the room owing to swell and decay is shown in Figure 1.3.18.

There is an optimum level which depends on the use of the hall: speech requires less than music. That of a room with a volume smaller than 100 m³ is about 0.9 s, whereas for a concert hall of 1000 m³, 1.3 s is considered ideal.

Reverberation can be limited by absorption, for instance, by curtains, which convert most of the sound energy into heat and movement. In a hall where the reverberation is too high for a given performance, closed curtains may have a beneficial effect. Audiences

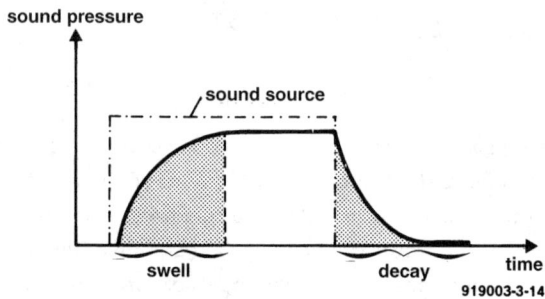

Figure 1.3.18. Pattern of sound pressure as a function of time at a given position in the room. The pressure rises gradually and slowly decays when the sound stops.

also cause absorption: it takes much experience to judge during a test without the public present how the sound will be when there is an audience. Also, open windows reduce the reverberation (but may cause a nuisance for the neighbours). In certain 'difficult' rooms (long ones with low ceilings and much glass), it may well prove sensible to cut out as much of the direct sound as possible by turning all the instrument loudspeakers, but not those of the voice amplifiers, towards a wall. The arrangement requires a lot of experimentation.

Another phenomenon that must be taken into account are air resonances, which have an important effect on the acoustics of a room. They occur when the wavelength of a sound is related by a whole number to one of the dimensions of the room. They manifest themselves by amplifying or attenuating certain frequencies wholly or partly. Since music contains an enormous number of wavelengths, air resonances cannot really be avoided. Which frequencies are affected depends entirely on the dimensions and arrangement of the room. Resonances are more or less pronounced at different locations in the room, which means that at these locations certain tones may dominate. Most people react subconsciously to resonance effects by slightly turning their head or assuming a different position. When recordings of a performance are to be made, microphones must not be placed in resonance locations: sometimes a position only a few centimetres away will improve matters appreciably.

1.3.5 Experiencing sound

The aim of making music is to make people enjoy the produced sounds. For this, it is not enough for the musician to play his instrument well. The sound must also be presented in an attractive and correct manner. This means that the sound engineer must know something of the human ear and hearing in order to understand many of the problems encountered in electro-acoustics. The human ear is a complex organ which converts vibrations of air molecules into nerve impulses that are processed in the brain. It ensures that ambient sounds are made audible, and at the same time it prevents sounds of the listener's body (heart beat, rushing of blood, intestinal rumblings) to be heard.

The vibrations of the air molecules are sensed by the ear drum (tympanic membrane). These fluctuations are passed on to the cochlea by three tiny bones or ossicles: hammer (malleus), anvil (incus) and stirrup (stapes). A lever mechanism, formed by the tympanic membrane and rigid-body rotational motion of the hammer and anvil, converts incoming vibrations with large amplitude but little pressure into vibrations with small amplitude but magnified pressure. When the amplitude of the vibrations becomes too large, the ossicles seize, which is clearly discernible and is called the threshold of pain.

The cochlea converts the vibrations into nerve impulses, which is effected by tiny hair receptor cells that convey information as to pitch and amplitude to the brain via aural nerves. When we grow older, the hair receptor cells gradually die, so that the sensitivity of the ear diminishes. This process is called presbyacusis. The sensitivity may also be impaired (and often is in even very young people) by persistent exposure to very loud sounds; this impairment is often called sociocusis. Short-term overloading of the ear does not necessarily lead to hearing impairment if the ear is given a long period of recovery. Long means at least ten times as long as the period of overloading. So, a rock concert lasting two hours should be followed by a rest period of at least 20, and preferably 36 hours. Rock concerts that last longer than four hours or where the sound pressure level consistently exceeds 120 dB or more lead with absolute certainty to permanent hearing impairment or, if repeated, traumatic deafness.

A young person can hear sounds at frequencies of 20–20,000 Hz (that is, the complete audio range). This hearing band slowly shrinks with age, particularly at the high end. In healthy people, hearing sensitivity at middle frequencies (2000–4000 Hz) changes little with age.

Sound just under the audible range (infrasound) is experienced as rumbling, while sounds above the audible range (ultrasound) are often detected as an indefinable pressure.

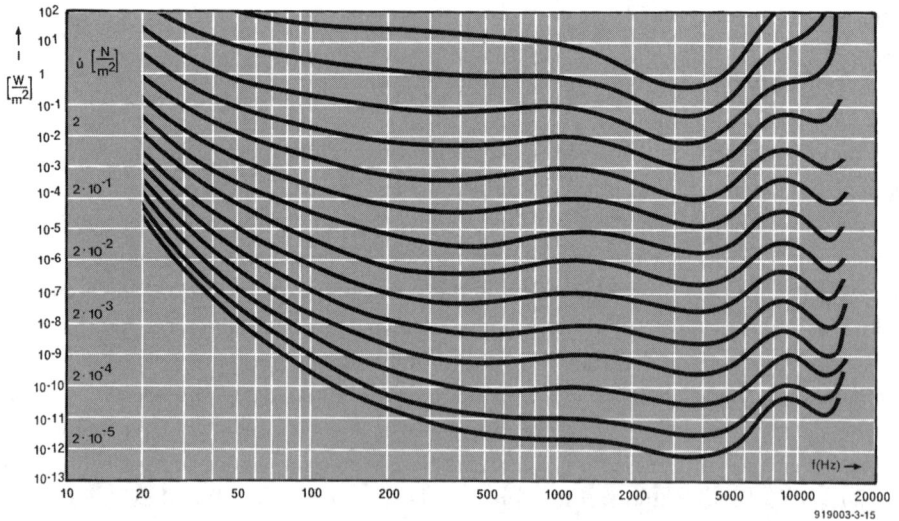

Figure 1.3.19. The characteristics make it possible for the functioning of the ear to be assessed. Loudness is experienced as equal along each of the curves. At small sound pressures, high and low frequencies must be radiated at a higher level than the middle frequencies. This sensation decreases with increasing sound pressure. The peak sensitivity of human hearing lies between 1000 Hz and 4000 Hz.

Both phenomena are unpleasant and must be avoided during music reproduction. Human hearing can process widely varying sound pressures: the lowest, called the threshold of hearing, depends on frequency. Our hearing is most sensitive at about 1000 Hz. At that frequency, the threshold of hearing is equivalent to a sound pressure, $p = 2 \times 10^{-5}$ N m^{-2}. The threshold of pain, at the same frequency, occurs at a pressure, $p = 110$ N m^{-2}. This vast range is not processed linearly, but logarithmically. This means that doubling the amplitude of a sound (corresponding to an increase in L_p of 6 dB) does not give the impression of a sound twice as loud. Doubling the sound power (that is, an increase in L_w of 3 dB) is just about discernible. It is therefore an erroneous assumption that doubling the power of an amplifier (with equal loudspeaker efficiency) leads to a doubling of the loudness.

The above is clarified in Figure 1.3.19, which shows a series of equal loudness contours or isophones. These show the amount of sound pressure required to produce sensations of equal loudness at various frequencies and volume levels. The contours indicate that the highest sensitivity of human hearing occurs at 1000–4000 Hz. This is why 1000 Hz is often taken as a reference frequency. To experience the same loudness level at lower or higher frequencies as at 1000 Hz, the amplitude of the signal must be increased appreciably. With increasing amplitude, this effect becxomes rather less marked. Many amplifiers therefore have a 'loudness' control, which arranges for the low and high frequencies to be amplified slightly more than the middle frequencies. This frequency characteristic must not be introduced during live performances.

Finally, it should not be overlooked that we have two ears and these enable us to hear in three dimensions or spatially, which makes it easy to localize sound sources. A live band performing without a PA installation forms a natural, spatial source of sound. It is important that this spatial impression is retained as much as possible.

1.4 Dynamic range

In simple terms, the idea of dynamics in electro-acoustics may be described as the difference between the lowest and highest signal levels. The signal level may be a sound pressure or an electrical voltage. The full dynamic range of our hearing can be determined in a spcial test room only. At the lower end the range is limited by the threshold of hearing (where a young, healthy person can just hear the sound) and at the upper end by the threshold of pain. As mentioned before, our ear can process sound pressures that differ by many powers of ten: at a frequency of 1000 Hz, this difference is 10^6 or 120 dB.

However, even in our domestic living room the dynamic range is limited by normal ambient noises. During live performances, the background noise is even greater. Playing

Instrument	Dynamic range
singing voice	about 60 dB
kettle drum	about 60 dB
clarinet	about 50 dB
grand piano	about 45 dB
snare drum	about 40 dB
acoustic guitar	about 40 dB
cymbals	about 35 dB
trumpet	about 30 dB

Table 1.4.1. Dynamic range of some musical instruments. Values shown are nominal.

loudly enables the music to be heard everywhere in the room or hall, but the loss in dynamic range is made greater by the logarithmic characteristic of our hearing.

When an amplifier system is used, there are even more limitations. The lower end of the dynamic range is limited by the noise emanating from the electronic circuits (noise floor) and the upper end by the maximum power of the system (overload). Recorded sounds are compressed, because there is not a sound installation that can cope with the original dynamic range. The result is that any recording compared with a live performance without amplifier sounds flat and loses much of its impression. This is particularly true of live music played on acoustic instruments.

Each process on an audio signal with the aid of electronic equipment limits the dynamic range of the original sound. This limitation is, in principle, irreversible, that is, the signal cannot be restored to its original quality by some technical trick. This means that the widest dynamic range can be attained only during a live performance with acoustic instruments. When instruments are amplified by electronic systems, a certain limitation of the dynamic range is unavoidable. In other words, the dynamic range of electric instruments (guitar, keyboards, percussion) is always smaller than that of acoustic instruments. Each and every further electronic process (for instance, reproduction by a PA installation, or the use of effects units) decreases the dynamic range even further. It is a fact that the more electronic links are used in the sound reproduction chain, the more the reproduced sound differs from the original. Less electronics means in many cases a greater dynamic range, more freedom of performance, and higher musical expression. This is why a rock band that uses electronic instruments predominantly cannot hope to equal the dynamic range of a classical orchestra. Playing extremely loudly only exacerbates the reduction in dynamic range. Each and every sound engineer should remember this. The dynamic range of a number of instruments is given in Table 1.4.1.

2. Microphones and musical instruments

This chapter deals with the actual production of musical sounds from the point of view of the sound engineer. Microphones and musical instruments are the signal sources that produce low-frequency fluctuations of air that in certain cases are processed by mixer panels and amplifiers. The principle is always the same: tones are generated by some mechanical construction (guitar, piano, voice, and so on), but before these can be processed further they have to be converted into electrical signals.

In general, components that convert sound energy into electrical energy, or vice versa, are called transducers. In electrical musical instruments, the mechanically generated vibration is normally converted directly into an electrical vibration. Acoustical instruments as well as the human voice in the first instance produce a mechanical vibration which is converted into an electrical vibration by means of a sound board or a microphone.

How these conversions take place and what aspects have to be minded is discussed in this chapter. It must be borne in mind that not every acoustic signal needs to be converted into an electrical signal: the limitations discussed in Section 1.4 continue to apply. Wherever possible, electronic signal processing is to be avoided.

2.1 Microphones

A microphone converts sound energy into electrical energy. The basic construction of all microphones is the same: a sensor detects vibrations of air molecules and converts them into mechnical vibrations, which are subsequently converted into electrical vibrations (Figure 2.1.1).

Depending on their application, converters are constructed in various manners and normally have very different properties. It is important to use a microphone that by its construction and properties is best suited to the application in hand. For this, knowledge of the function and construction of the major types of microphone is, of course, indispensable. As has been mentioned before, during each and every process the signal undergoes, a small part of the original is lost. This is also true of a microphone: in principle, it is impossible to obtain a spatial impression of a sound via a microphone: after all, the sound is 'hearsd' at only one location at a time. This unfortunate aspect can be alleviated to some extent by the use of a number of microphones (see Section 6.1). Note that there is also some loss of frequency range and dynamics.

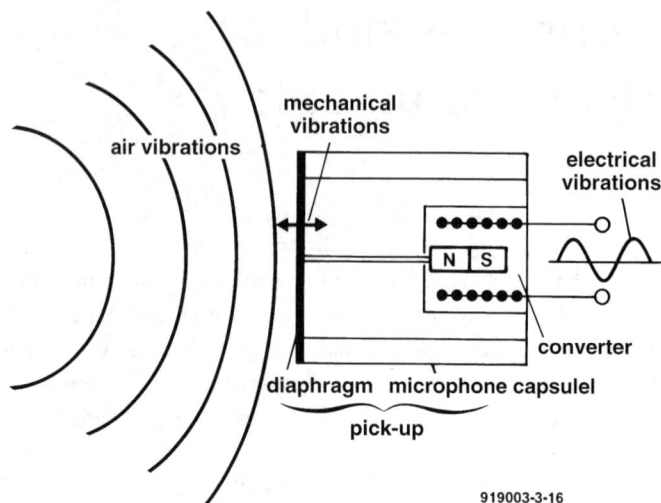

Figure 2.1.1. Basic construction of a microphone. The vibrating air particles are converted into mechanical vibrations by a diaphragm. The converter transforms the movements of the diaphragm into electrical voltages. Sound energy is therefore converted twice to a different kind of energy.

2.1.1 Pickups

The pickup of a microphone converts vibrations of air (sound) into mechanical vibrations. The directivity, the performance near a sound source and, to some extent, the frequency characteristics of a microphone are determined by the mechanical construction of the pickup. There are two basic but different constructions: the pressure pickup and the pressure gradient pickup.

A schematic representation of a pressure pickup is shown in Figure 2.1.2. The cone or diaphragm is stretched across a shell whose rear has a pressure compensation aperture. The dimensions of the aperture ensure that the air pressure in the shell is the same as that outside, but rapid changes in air pressure caused by sound waves are not compensated. The diaphragm therefore moves when the outside pressure is different from that inside the shell. Sound waves from the rear or the sides of the shell also cause differences in pressure inside and outside the shell so that the cone moves. In other words, the pickup reacts to sound waves from all directions and is thus omni-directional. This is true for frequencies up to about 1000 Hz. Diffraction decreases as rhe wavelength approaches the diameter of the diaphragm, so the response to off-axis sounds diminishes at higher frequencies. The microphone thus becomes increasingly directional as the frequency rises; the smaller it is, the higher the starting point. The typical polar

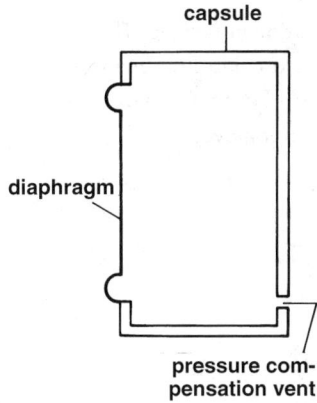

capsule

diaphragm

pressure com-
pensation vent

919003-3-17

Figure 2.1.2. Principle of a pressure pick-up. The diaphragm reacts to sound from all directions without any preference.

diagram of a pressure pickup is shown in Figure 2.1.3. For an acceptable sensitivity, the diaphragm of the microphone must have a diameter of not less than 15 mm. This corresponds to a quarter of the wavelength of an air vibration at 5.7 kHz. This means that at frequencies above about 5.7 kHz the microphone becomes more and more directional.

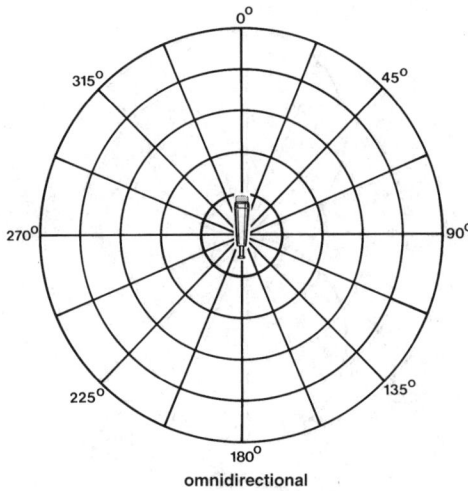

0°

315°   45°

270°   90°

225°   135°

180°
omnidirectional
919003-3-18

Figure 2.1.3. Polar diagram of a pressure pick-up: its sensitivity is equal in all directions.

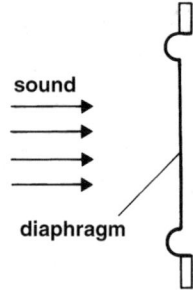

Figure 2.1.4. Principle of a pressure-gradient pick-up. The diaphragm reacts only to sound waves arriving from the front or the back. Sound waves at the sides produce equal air pressures so that the diaphragm does not move.

919003-3-19

If the back of the cone is open to the air as as in a pressure gradient (velocity or ribbon) microphone, shown schematically in Figure 2.1.4, equal pressure on both sides of the diaphragm results in zero movement. Front-propagated as well as rear-propagated sound waves cause the diaphragm to deflect, whereas those from either side have hardly any effect. This means that a pressure gradient microphone has a polar diagram shaped like a figure-of-eight as in Figure 2.1.5.

In contrast to the pressure microphone, the pressure gradient type reacts to (air) particle velocity. In the immediate vicinity of the sound source, this velocity increases rapidly for low frequencies. This causes a pressure gradient microphone to reproduce low frequencies much louder than low and middle frequencies (proximity effect). Pads are sometimes inserted behind the

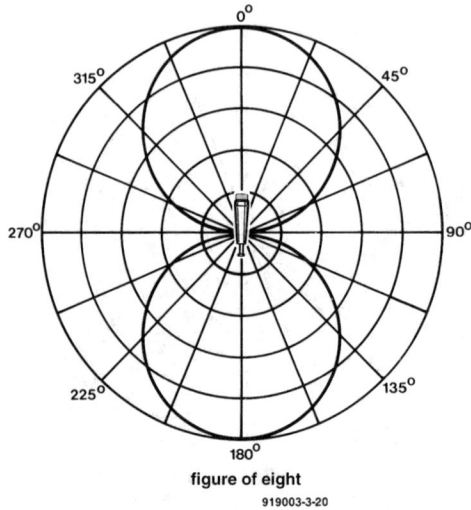

figure of eight

919003-3-20

Figure 2.1.5. The polar diagram of a pressure-gradient pick-up is a figure of eight.

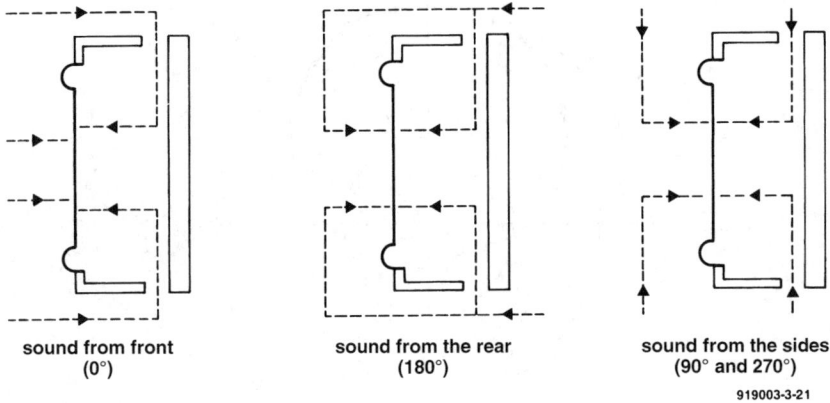

| sound from front (0°) | sound from the rear (180°) | sound from the sides (90° and 270°) |

919003-3-21

Figure 2.1.6. Operation of an acoustic filter (delay element): the difference in the transfer times between the sound waves arriving at the front and those arriving at the back depends on the angle of incidence of the sound. When the sound arrives at right angles (0°), the sound waves at the back are delayed with respect to the sound at the front in proportion to the distance travelled. The consequent pressure difference causes the diaphragm to move. When the sound arrives at the back (180°), the transfer times of the waves at the front and those at the back are about equal. There is then no difference in air pressure so that the diaphragm does not move. The difference in transfer times of sound waves arriving at the sides (90° and 270°) is so small that the diphragm does not move.

diaphragm to restrict rear response. If this reduces to just a small rear lobe, the polar diagram becomes a cardioid (Figure 2.1.7), a supercardioid (Figure 2.1.8), or a hypercardioid (Figure 2.1.9).

There is a difference in transfer times for front-propagated waves to reach the front and rear of the diaphragm. The waves reaching the rear have travelled a longer distance than those incident on the front, and they therefore arrive a little later. The resulting pressure difference leads to a deflection of the diaphragm. When the sound source is at the back of the microphone, the transfer times to front and rear are about the same. There is then no pressure difference and the diaphragm does not deflect. Off-axis sounds do result in a transfer time difference. This is small, however, when the sound is propagated directly at the front, and the corresponding movement of the diaphragm is correspondingly tiny. When the sound source is moved from in front of the microphone to at the back of it, the difference in transfer time s becomes smaller, and the pressure difference and diaphragm deflection become correspondingly smaller also. Depending on the construction of the acoustic pad, a number of different polar

71

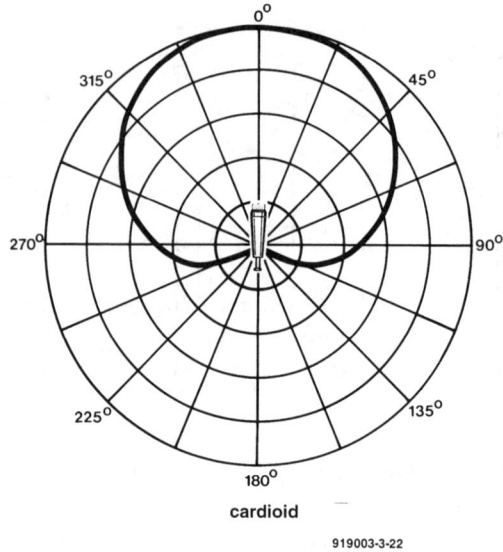

cardioid

919003-3-22

Figure 2.1.7. The polar diagram of a pressure-gradient pick-up is cardioid-shaped.

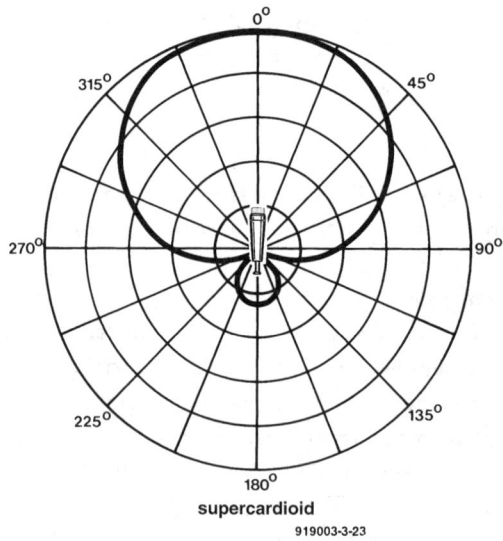

supercardioid

919003-3-23

Figure 2.1.8. The polar diagram of a pressure-gradient pick-up is altered to a super-cardioid shape with the aid of acoustic filters.

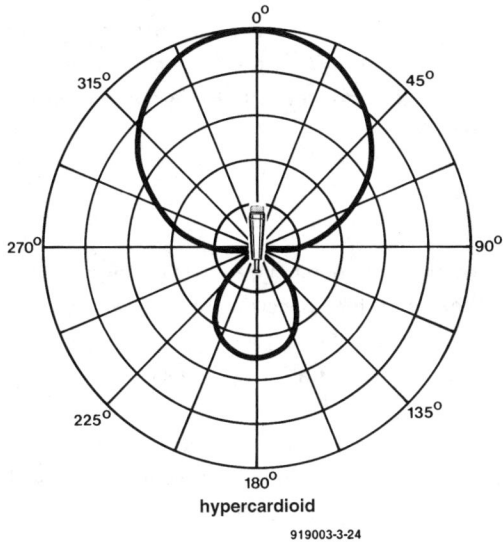

Figure 2.1.9. The polar diagram of a pressure-gradient pick-up fitted with a special acoustic filter is hyper-cardioid-shaped. The diagram shows that when the forward sensitivity of the microphone increases, it also becomes more sensitive to sounds from other directions (here, to sounds from the back), which gives rise to side lobes.

diagrams can be realized in this manner.

With constant sound pressure, a microphone with a cardioid polar diagram may be placed 1.7 times, one with a supercardioid polar diagram, 1.9 times, and one with a hypercardioid polar diagram, 2 times, further from a sound source than an omni-directional microphone to produce the same signal level. Note that microphones with a supercardioid or hypercardioid polar diagram are more sensitive to sound waves coming from the rear.

It will be clear that the more sensitive a microphone is to front-propagated sound, more spurious side lobes arise, normally at the rear of the microphone. This obviously increases the sensitivity of the microphone to sounds from the sides or rear, which may be extremely disturbing in practice. Some manufacturers do not use acoustic resistance pads . Directivity is then obtained by the use of combining different types of shell in one. This makes no difference whatsoever to the user.

Apart from the directivity, there are other important parameters. To obtain as high a signal-to-noise ratio as feasible, a high signal level is imperative. This means that the diaphragm must be able to deflect strongly, but this may adversely affect the frequency response and cause distortion. These are contradictory requirements, which make it clear that as so often a compromise has be found. In one application, a good signal-to-noise ratio may be the most important requirement, and in another a good linear frequency response. For high sensitivity, the diaphragm must be light and very mobile. At the same time, it must be able to handle large sound pressures (up to 140 dB) and this means that it needs to be mechanically robust. Again, the designer has to compromise.

2.1.2 Transducers

There are several ways of converting sound into electrical energy and this the task of the electro-acoustic transducer. Since most transducers can be combined with most pickups, microphones are classified in accordance with the transducer: carbon microphones, ribbon microphones, moving-coil microphones, crystal/ceramic microphones, capacitor microphones, and others.

In the simplest type of microphone, the carbon microphone, diaphragm movements are converted into electrical signals by carbon granules that are compressed to a lesser or greater degree – depending on the sound pressure – by the diaphragm (see Figure 2.1.10). The resistance of the mass of carbon granules depends on the pressure, so that, when a voltage is applied across the microphone, the current will vary in rhythm with the sound waves in accordance with Ohms's law. Note that carbon microphones need a supply voltage, which is applied via the normal terminals. The properties of a carbon microphone are such that these microphones are really only suitable for spoken information. They are used in sound engineering in combination with headphones for communication between the musicians and the mixer operator. Since carbon microphones are still used in hundreds of m illions of telephones, it is even today the most widely used type.

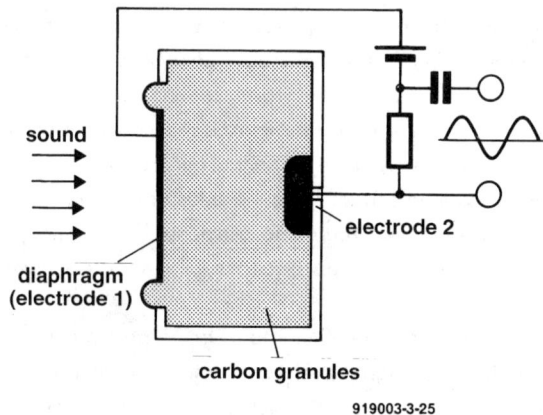

919003-3-25

Figure 2.1.10. Principle of a carbon microphone.

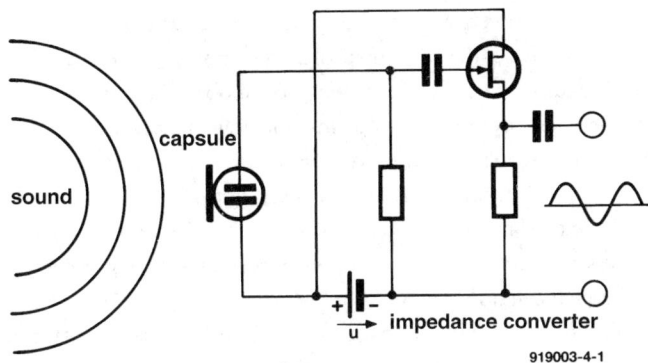

Figure 2.1.11. Principle of a capacitor (condenser) microphone.

The capacitor microphone has become the standard one for all sorts of sound recording. In this type of microphone, the diaphragm and second, fixed, electode form a capacitor whose capacitance varies in line with the deflection of the diaphragm. Such a change in capacitance is easily converted into an audio-frequency signal. The principle of this type of microphone is shown in Figure 1.2.11.

The diaphragm consists of a very thin metal film that weighs only a few grams, which largely contributes to the excellent properties of the capacitor microphone. A capacitor microphone contains an integral preamplifier that matches the high internal resistance of the microphone to the input impedance of the mixer or tape recorder. The circuit of the preamplifier does, of course, contribute to the properties of the microphone, particularly as regards the noise level and the dynamics. The capacitor microphone also needs a supply voltage, partly for the preamplifier and partly to sustain the charge of the capacitance. The supply voltage is normally provided by the mixer or the tape/cassette recorder.

A new form of capacitor microphone is the electret capacitor microphone, normally just called electret microphone. This type of microphone has a fixed charge for which it needs a small, 1.5 V integral battery to power the internal preamplifier.

Properties common to all capacitor microphones are the high sensitivity, the linear frequency response and the excellent impulse performance.

One of the most frequently used microphones on the amateur stage is the moving-coil type. Its principle of operation is analogous to inductively generating a voltage (1.1.1). The diaphragm is permanently fixed to a small coil or an aluminium ribbon, which is located in the field of a permanent magnet. Movements of the diaphragm alter the strength of the field surrounding the coil, which causes small low-frequency alternating voltages to be induced in the coil. The weight of the diaphragm is about twenty times that of a capacitor microphone. Consequently, its impulse performance and sensitivity are nowhere near as good. On the other hand, this type of microphone does not need an integral power supply, which makes its use rather simpler. Moreover, its dyamic range is larger owing to the absence of a preamplifier.

Since moving-coil microphones work on the principle of induction, trouble may be experienced with stray fields, resulting in annoying hum. Such stray fields are caused by electrical apparatus near the microphone, such as amplifiers, lights, dimmers, and so on. Some moving-coil microphones have in series with the pickup coil an integral compensation coil , which also picks up the stray field. Since the compensation coil is wound in the opposite direction from the pickup coil, the voltages induced in the two windings largely nullify one another.

The operation of crystal/ceramic microphones depends on the piezoelectric effect of a crystal. Firmly secured at one end, the crystal generates an electrical signal when it is stressed by a vibrating cone in contact with the other. Since natural crystals such as rochell salt and quartz are fragile and adversely affected by humidty and temperature, more stable synthetic substances such as barium titanate and lead zirconate are used in their place. A crystal microphone is much less expensive than a capacitor microphone. Nevertheless, owing to its sensitivity to humidity and temperature changes, this type of microphone is not often used on the stage or in the studio. Piezoelectric transducers are, however, frequently used as pickup in instruments.

2.1.3. Types, properties and applications

The pickup and transducer of a microphone form a vibrating system which, of course, has a self-resonance frequency. When the microphone is used with audio signals near this frequency, the induced voltage increases greatly, and the frequency response becomes distorted to a greater or lesser erxtent. It is as if near-resonant frequencies are treated preferentially.

The resonant frequency (tuning) of a microphone is determined during manufacture, in which shape and dimensions are of great importance. Designers endeavour to avoid sharp resonance peaks, but, in spite of this, the peak is clearly recognizable in the response curve of each and every mikcrophone. The manner of tuning indicates for which applications a microphone is or is not suitable. For instance, the resonant frequency of a moving-coil micropone with pressure gradient pickup is low, whereas with a pressure pickup, it is mid-range; that in a capacitor microphone with a pressure-gradient pickup is mid-range, and with a pressure pickup, it is high. This means that, in general, the resonant frequency of capacitor microphones is higher than that of moving-coil microphones. Microphones with low resonant frequencies are more sensitive to contact and cable noise, and to wind noise. Those with high resonant frequencies have a tendency to howl. This kind of positive feedback results from the microphone picking up part of the sound from the loudspeaker and feeding it back via the amplifier.

In view of the required immunity from ambient noise, microphones used on the stage and in the studio are invariably low-impedance (low Z) types, and are, therefore, called low-impedance microphones.

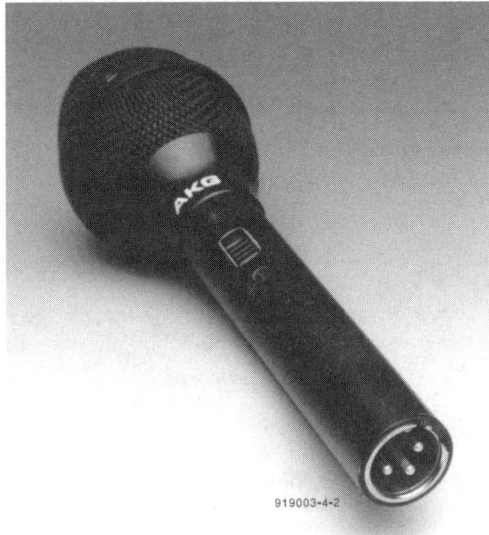

Figure 2.1.12. Recessed 3-pin (male) XLR microphone connector.

A professional microphone is invariably terminated into a 3-pin XLR plug with recessed pins as shown in Figure 2.1.12. The pin connections are shown in Figure 2.1.13. Pin 1 is connected to the housing (shielding), whereas pins 2 and 3 are linked to the pickup coil. In low-impedance microphones, Z_{micr} is between 150 Ω and 600 Ω. Therefore, the input impedance, Z_{in}, of the equipment (mixer, voice amplifier) to which the microphone is to be linked must be not lower than 600 Ω. This will normally be the case in modern equipment. An excellent situation occurs when the input impedance of the mixer is slightly higher than the impedance of the microphone. Some possible combinations are:

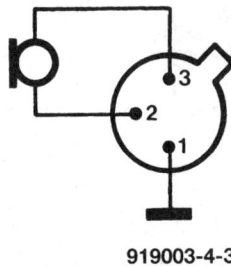

919003-4-3

Figure 2.1.3. Pinout of an XLR microphone connector.

77

Z_{micr}	Z_{in}	comment
high	high	loss of high frequencies when connecting cables are long
high	low	does not work; a transformer needs to be used
low	high	signal attenuated by about 20 dB
low	low	best combination; even long cables may be used without losses

Table 2.1.1.

Large mismatches can be corrected with a suitable microphone transformer. The circuit diagram of such a transformer is shown in Figure 2.1.14. The high-impedance component must be connected across the high-impedance winding, and the low-impedance component across the low-impedance winding. In the diagram, the output of a low-Z microphone is applied to a high-Z amplifier input via a 1:100 microphone transformer. In such a setup, it must be ensured that the signal transmitted via the long microphone cable is low-impedance. In other words, the cable must always be connected across the low-Z transformer winding.

If the connecting cables are not longer than about 10 metres (33 ft), the microphones may be connected unbalanced by a single, screened cable as shown in Figure 2.1.15 (see also Figure 2.1.16). The cable must be terminated at one end into an XLR socket (see Figure 2.1.17) and

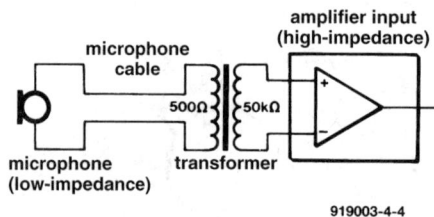

Figure 2.1.14. Connection diagram of a low-impedance, say, 300 Ω, micophone to a high-impedance input of an amplifier or mixer. The transformer must be connected in such a way that the microphone cable always carries the low-impedance signal. In this diagram, the transformer should preferably be built in the amplifier.

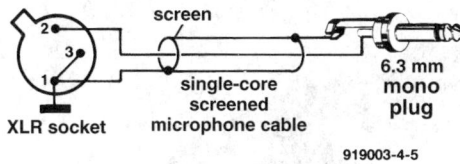

Figure 2.1.15. Connection of an unbalanced (asymmetrical) microphone cable.

at the other end into a 6.3 mm mono plug (see Figure 2.1.18). Pins 1 and 3 are interlinked and soldered to the cable screen. Pin 2 is the signal (hot) line. Unfortunately, this type of connection is relatively susceptible to noise and interference since the cable functions as an antenna that can and does pick up stray fields (in some cases even broadcast transmitters).

Much better results, at least as far as immunity to noise and interference is concerned, are obtained with a balanced connection, which uses a two-core screened cable as in Figure 2.1.19. Such cables may be up to 600 metres (almost 2000 ft) long without loss of quality. A condition is, however, that the amplifier or mixer is suited to such a connection. Here again, the cable is terminated into an XLR socket at one end an a 6.3 mm audio plug at the other.

As we have seen, capacitor microphones need a supply voltage. In the case of electret microphones this is not much of a problem since they can do with a small battery. Standard capacitor microphones, however, need a supply voltage of 12–48 V. There are two way of supplying this voltage to the microphone without adding to the number of cores in the cable: phantom

Figure 2.1.16. Microphone with unbalanced (screened single-core) cable.
The cable has an XLR connector at one end for accepting
the microphone and a 6.3 mm audio plug at the other end.

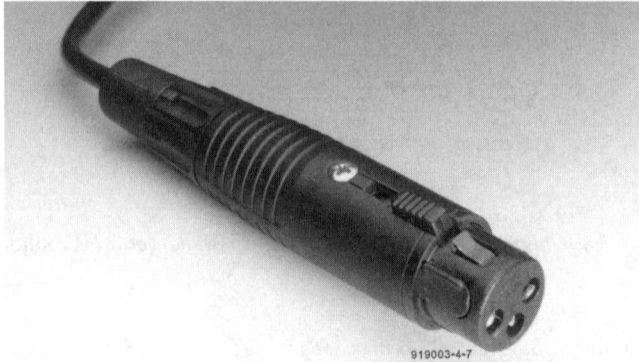

Figure 2.1.17. Three-pin in-line female XLR connector.

Figure 2.1.18. 6.3 mm audio plug.

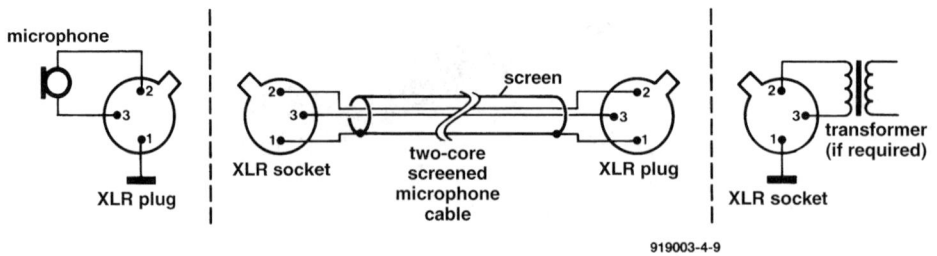

Figure 2.1.19. Balanced microphone connection. The screen is connected independently to pin 1. This arrangement makes possible cable lengths of up to 600 metres.

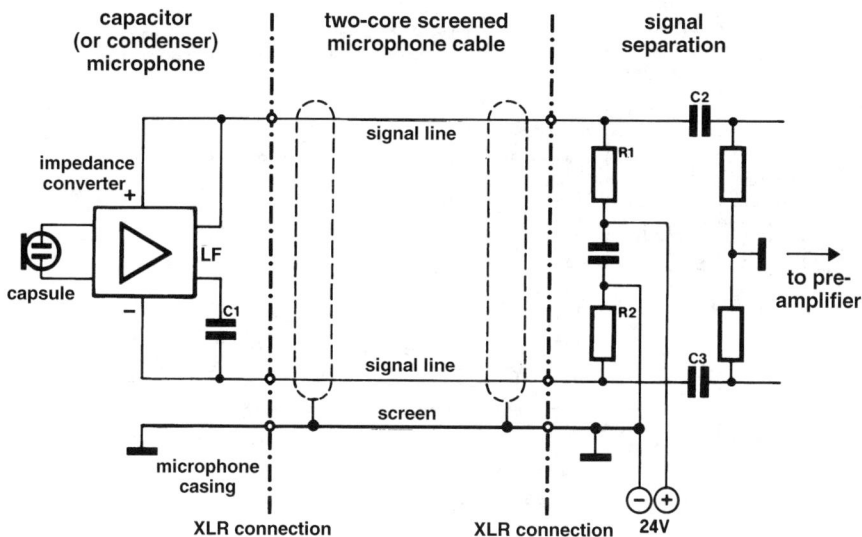

Figure 2.1.20. *Principle of supplying power to a capacitor microphone via the signal lines.*

supply and supply via the signal lines.

Supply via the signal lines is possible with balanced as well as unbalancec connections as described in DIN 45595. A basic diagram of such a connection is shown in Figure 2.1.20. In this diagram, the supply voltage is applied via the two signal lines. One of these lines is linked to the +ve supply via resistor R_1 and the other, together with the screen, to the −ve supply via R_2. Capacitors C_1 and C_3 isolate the supply lines from the input of the preamplifier. In this setup, an unbalanced connector may also be used. Bear in mind that moving-coil microphones do not need a supply voltage and they should, therefore, be linked to the signal-cum-supply lines when the power supply is switched off. If this is not so, a direct current will flow through the pick-up coil of the microphone which will lead to distortion of the sound and overheating of the coil, which may cause irreparable damage. Owing to these risks, this type of supply is not often used nowadays, since phantom supplies and balanced connections have been found to be better, more reliable, and not so risky.

A basic setup of a phantom supply is shown in Figure 2.1.21 − note that this can only be used with balanced connections. The supply voltage is applied to the signal lines via resistors R_3 and R_4. The cable screen provides the return path. This way of providing a supply voltage is described in DIN 45596. Since there is no direct voltage differential between the two signal lines, moving-coil microphones may be connected without the need of switching off the supply voltage. Unbalanced microphones cannot be used, however, since there would be a direct current through the pickup coil, which would almost certainly be destroyed.

81

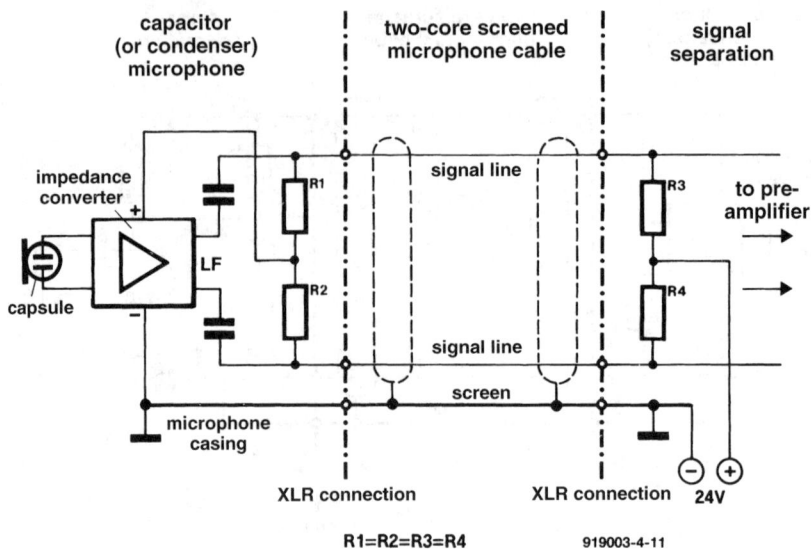

Figure 2.1.21. Basic diagram of a phantom power supply.

In modern mixers and good-quality voice amplifiers, a phantom supply is now built in as a matter of course. Such a supply is also available as a stand-alone unit for connection between the input of the amplifier or mixer and the microphone as shown in Figure 2.1.22. A phantom supply can also be built at home or in a small workshop: a suitable circuit diagram is shown in Figure 2.1.23. Note that for satisfactory operation, the tolerance of the resistors must be at least as shown or better.

Capacitor microphones may be used anywhere where good, natural sound reproduction is required and the sound level is not very high. Examples are good-quality tape recordings and recording of acoustic musical instruments with low to medium sound level, such as cymbals and acoustic guitars, recorder, and others. Some modern models may also be used for voice purpose. However, their use as voice microphone on the stage is not advisable since owing to their high sensitivity and high resonant frequency they have a tendency to howl. Moreover, they cannot cope with high sound pressures.

Moving-coil microphones are particularly robust and may be used for all kinds of suitable task on the stage. They are capable of working with high sound pressure levels (depending on type and make up to 140 dB!). Their frequency response is not as straight as that of a capacitor microphone, which leads to a slight colouring. Clearly, the right microphone must be used for each and every application .

Microphones that have a pronounced proximity effect are very suitable as voice microphone

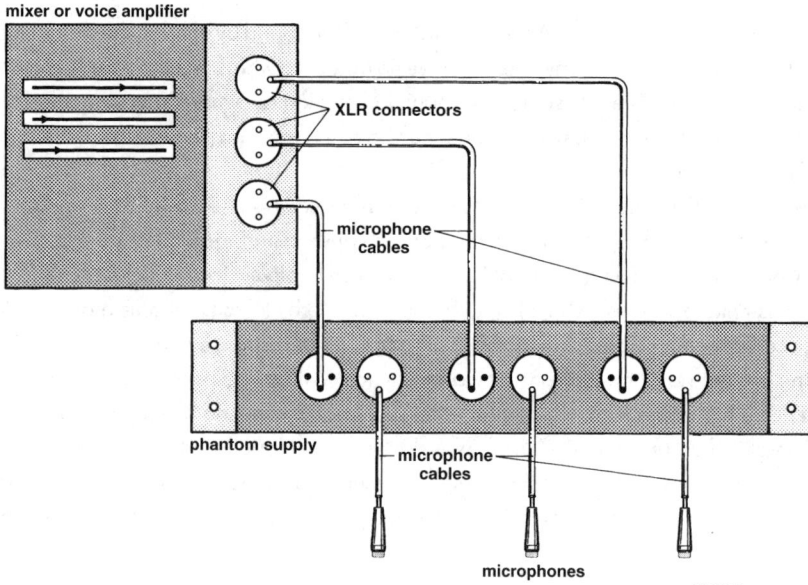

Figure 2.1.22. A good way of connecting a discrete phantom power supply.

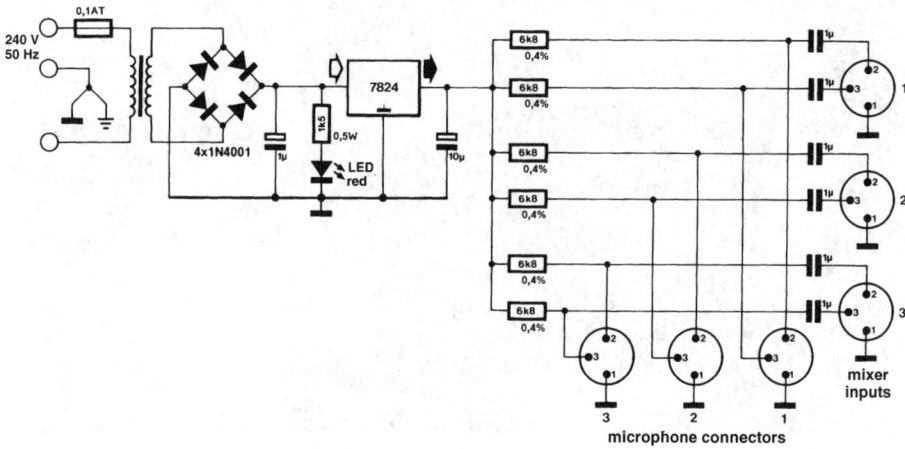

Figure 2.1.23. Circuit diagram of a phantom power supply. Such a small circuit is easily constructed on a small piece of prototyping board and built into, say, a mixer.

since the vocalist can vary the timbre and volume of the sound by altering the distance between mouth and microphone. Many listeners have no objection whatsoever to a frequency response that is not straight – after all, subjective sound impressions are more important than objective, technical criteria. The voice microphone thus has an important part to play in the final sound quality of the voice.

When the microphone is held in the wind or is moved up and down, or to and fro, rapidly, wind noise is generated. Such air disturbances cause disturbing and annoying rushing noises, which can be reduced by up to 20 dB by the use of plastic foam shields. When the microphone is held too close to the mouth, the noise made by the consonants b or p, called popping, can be reduced by a similar shield. The frequency response and directivity of the microphone are not noticeably affected by such shields. With some microphones, it is possible to switch in a high-pass filter, which attenuates low frequencies and rushing and popping.

There is a vast number of microphones on the market for all kinds of application. Dramatic improvements in their basic design are not foreseen: most manufacturers produce new types along the same lines as proven existing designs. Price is becoming more important, however.

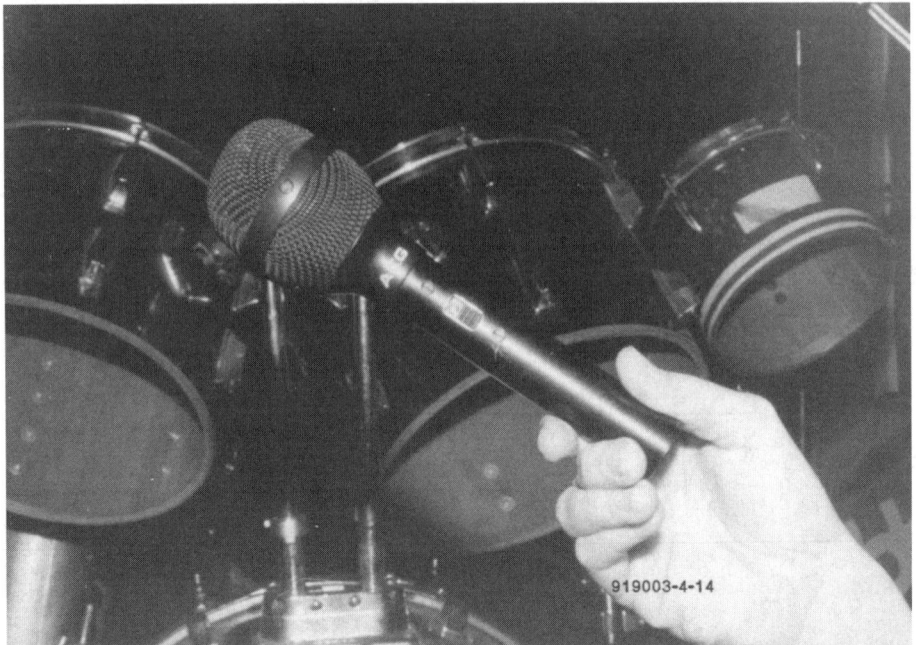

Figure 2.1.24. Electrodynamic microphone Type AKG D-80. This type of microphone is very popular with amateur bands because of its good electro-acoustic properties and its reasonable price. It is eminently suitable for voice, but also for recording drums.

84

Most manufacturers aim at marketing new models in the medium price range of £50–100. These models are invariably of excellent quality.

Dynamic microphones with a cardioid polar diagram are ideal for voice recordings. Their directivity and the tough membrane (compared with that of a condenser microphone) ensure that there is little risk of howl. Greater directivity would cause the instruments behind the singer(s) to be included. The D-80 from AKG (Figure 2.1.24) is a typical voice microphone in the economy price range. This microphone has most of the good characteristics of higher priced models, but no facilities to prevent contact and other mechanical extraneous noise. The SM58 from Shure is somewhat of a standard for rock and pop vocalists.

Voice microphones must not only have good electroacoustic characteristics, but also be mechanically robust (for example, it must be able to withstand a drop from the stand without getting damaged).

Special applications require special microphones. Such microphones are designed for just one application and are invariably not suitable for other purposes. The most important are wireless microphones, boundary microphones, gun microphones, and pickup microphones,

Wireless microphones, familar to most of us through television, are fitted with a miniature radio transmitter and are attached to actors' or speakers' clothing to obviate the use of microphone cables. Owing to their frequency characteristic, they are suitable for speech only (discussions, talk shows, and so on).

Gun microphones are highly directive owing to an interference tube fitted in the capsule. The directivity effect is obtained through the front waves arriving at the diaphragm in phase, whereas the out-of-phase lateral waves in the tube cancel each other out. These microphones are excellent for use out of doors, because they can pick up a particular sound from among a lot of ambient noise.

Boundary microphones, often called by their proprietary name PZM (=pressure zone microphone), are eminently suitable for tape recording music groups. When a recording is made with standard microphones, the reproduction often sounds hollow and only vaguely reminiscent of the original sound. This is caused by the reflections of the sound waves from the walls and ceiling of the recording room. These reflected sound waves arrive at the microphone a little later than the direct sound waves. This causes frequency-dependent amplification or attenuation of the sound (see also 1.3.4). Experimental research has shown that these phenomena do not appear near smooth surfaces. When PZM microphones are placed on a smooth surface a hemispherical polar pattern ensues. The sound so recorded is particularly differentiated and well-balanced.

Another special microphone is the crystal microphone (which depends on the piezo electric effect – see 1.1.1) specially intended for recording acoustic instruments and which, therefore, are often fixed to the sound board of a guitar. This enables the musician to move around more freely than when he/she is playing in front of a microphone. The microphone then does not pick up air vibrations, but rather the vibrations of the sound board. These vibrations have different properties from air vibrations and this is the reason that instruments so recorded often sound quite different.

To judge the usefulness of a microphone, we need its technical data. The significance of these will be discussed on the basis of the data of a Beyer dynamic microphone Type M400N in Table 2.1.2.

Type	Pressure gradient pick-up
frequency range	50–15000 Hz
polar diagram	super-cardioid
directivity	20 dB at 135°
free-field sensitivity	2 mV/Pa at 1 kHz
sensitivity to magnetic fields	4 μV/5 μTesla (50 Hz)
electrical impedance	500 Ω
input impedance	1000 Ω
diaphragm	polycarbonate
housing	body: aluminium; head: steel
finish	body anodized matt black
dimensions	diameter of body: 24/32 mm
	diameter of capsule: 51 mm
weight	258 grams
connector	3-pin Canon (XLR)

Table 2.1.2

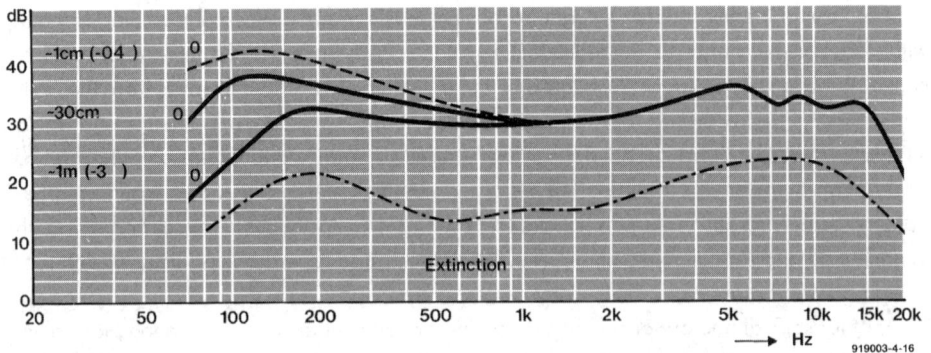

Figure 2.1.25. Typical frequency characteristic of voice microphone Type AKG D310. When it is held close to the mouth, low frequencies are emphasized. The dot-dash line is the characteristic curve of sound reaching the microphone from the rear.

The frequency range indicates the band of frequencies that can be handled by the microphone. Strictly speaking, these figures by themselves are not wholly sufficient, because within the range certain bands may be reproduced at a lower or higher level than adjacent bands. Moreover, frequencies outside the range stated are also (at least partly) reproduced. Clearer information on the frequency range is shown by a characteristic curve as in Figure 2.1.25.

The frequency response of a microphone must be suitable for the desired application. For instance, it is disturbing when the lowest and highest frequencies are reproduced by a voice microphone: after all, the human voice cannot produce these frequencies. Too wide a frequency range may mean that the double bass and cymbals are also picked up by the microphone and reproduced by the voice amplifiers.

The directivity indicates how sensitive the microphone is to sound coming from its sides. In the table, a frontal signal is reproduced at least ×10 stronger (20 dB – Eq. 21, p. 38) than a signal arriving at the microphone at an angle of 135°. This parameter is not always stated in manufacturers's specifications and is in any case not required when the polar diagram is given or the microphone is an omnidirectional type.

The free-field sensitivity is a measure of the sensitivity of the microphone capsule. The M400N produces an output of 2 mV when the sound pressure level is 1 Pa at 1000 Hz.

The sensitivity to magnetic fields indicates the spurious output level of the microphone in a stray magnetic field. In this case, the spurious output is 4 μV when the flux density of the magnetic field is 5 μT (T = Tesla which is 1 weber per m^2). The lower this sensitivity, the better the microphone is protected against magnetic interference (for instance, by a compensating inductor). When such protection is not specifically provided, the parameter is not normally mentioned in manufacturers' specifications.

The electrical impedance is the internal impedance of the microphone.

The input impedance is the minimum input impedance of the amplifier or mixer the microphone is linked to. If it is connected to a lower input impedance, its performance may be degraded.

The other parameters in the table refer to physical parameters which do not need a further explanation.

Apart from the parameters in Table 2.1.2, the maximum sound pressure is often stated. When the actual sound pressure is greater than this value, the distortion increases to more than one per cent. In the case of a dynamic microphone, the maximum sound pressure is normally much higher than pressures encountered in practice, so that the parameter is normally not stated. In the case of a condenser microphone, however, the maximum allowable pressure is much lower and should therefore be stated in its data sheet. It is especially important to know the maximum permissible sound pressure when the microphone is used with loud instruments like a bass drum.

2.1.4 Some practical hints

It is clear from the foregoing that a voice microphone should not pick up the sounds from musical instruments. Unfortunately, this cannot always be avoided in practice since these instruments are normally placed behind the vocalist(s). Even in a rehearsal room it cannot be prevented owing to lack of space. This means that particularly in the case of a small stage, the singer(s) must hold back to some extent. In the rehearsal room, the voice microphone is placed as far away from the musical instruments as possible and, if possible, in such a way that all that is behind the vocalist(s) is a sound-absorbing wall.

There is also a technical trick which reduces the level at which sound from musical instruments is picked up—see Figure 2.1.26. It makes use of two microphones mounted in close proximity and interconnected in anti-phase. Both microphones pick up the same background sounds. These sounds will produce the same output level from the microphones but, since they are in anti-phase, will largely nullify each other. The vocalist(s) use sonly one of the microphones. This will therefore produce a higher output than the other, so that in the final analysis only the voice signal less the background sounds is reproduced and amplified. Although the microphone which is not used by the vocalist(s) need to be only a small, inexpensive type, some parameters of the two microphones, particularly directivity and sensitivity, must be (near) identical.

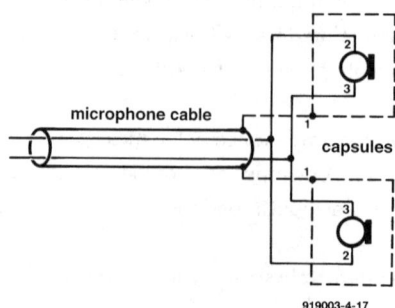

919003-4-17

Figure 2.1.26. Two microphones connected in series anti-phase
effectively suppress much of the background sounds.

Omni-directional electret microphones are eminently suitable for simple tape recordings. Their location in the rehearsal room must be chosen carefully to ensure that all musical instruments and vocalist(s) are recorded in correct proportion. It is particularly important that no microphone is located at a point of resonance. It is always advisable to make some test recordings before the final recording session.

The correct choice of microphone is also of great importance when the signal is to be used with a public-address (PA) system. There are special types of microphone that because of their frequency response (emphasis on low frequencies) and construction are particularly suitable for recording double bass, bass drums and low-frequency wind instruments (tuba). There are miniature microphones available that may be fitted on or in certain musical instruments. This makes the setting up of the other microphones much easier and it also provides more space on the stage. These microphones are invariably condenser types that need a phantom supply.

2.2 Musical instruments

In sound engineering, musical instruments are graded into acoustic types, electric types and hybrid types (part acoustic, part electric).

Acoustic instruments, for instance, guitars, pianos, violins, saxophones, and timpani have no electrical or electronic parts. The tone is generated by mechanical vibrations (vibrating string or taut skin) and amplified by a sound box or board. Resonance only occurs when the ratio of the dimensions of the sound box and the generated vibration is a whole number. Since the wavelength is inversely proportional to the frequency, the sound box of bass instruments must be appreciably larger than that of a guitar or violin. This is why bass guitars with a small sound board (acoustic bass guitar) can generate harmonics only. Only the double bass has a sound box that is large enough to produce the fundamental frequency also.

As already mentioned in section 1.4, the good properties of a musical instrument are degraded by electrical magnification. This is the reason that acoustic instruments are picked up by a microphone and subsequently amplified only when this is imperative as, for instance, when the sound is not strong enough for the volume of the hall. However, since in pop and rock music performances microphones are used for almost any instrument, the following sections will discuss how the various musical instruments are best magnified with the aid of a microphone and amplifier.

Electrical instruments such as the electric guitar and electric keyboard have no sound board at all: they merely produce electrical signals. An amplifier and loudspeaker(s) take the place of the sound board.

Hybrid instruments consist of acoustic instruments with built-in pickup (Ovation guitars, for instance) and electrical instruments with an additional sound board (Gibson ES335 guitar, for example). The concepts for amplifying the sound of electric and acoustic instruments are identical.

2.2.1 Guitars

The sound of a guitar is produced by a vibrating string. The modern guitar, shown in Figure

2.2.1, has six strings, is about 65 cm long, and is tuned to E_2, A_2, D_3, G_3, B_3, and E_4 (corresponding to frequencies, f_o, of 82, 110, 147, 196, 247, and 330 Hz). The fundamental frequency f_o at which a string oscillates is determined by the length l, mass m, and the tension T of the string:

$$f_o = \{\sqrt{(Tl/m)}\}/2l \qquad\qquad \text{[Eq. 30]}$$

Apart from the fundamental frequency, a large number of overtones are produced whose frequency is a whole multiple of the fundamental—see Figure 2.2.2.

The sound colouration of a guitar is determined primarily by the ratio of the amplitudes of the fundamental and the individual overtones. The ratio itself depends on the manner and force with which the string is plucked (with or without a plectrum), the position at which the string is plucked, the kind of string, and the construction of the sound board. Equation 30 may be rearranged in terms of string tension T:

$$T = (m/l)(2fl)^2 \qquad\qquad \text{[Eq.31]}$$

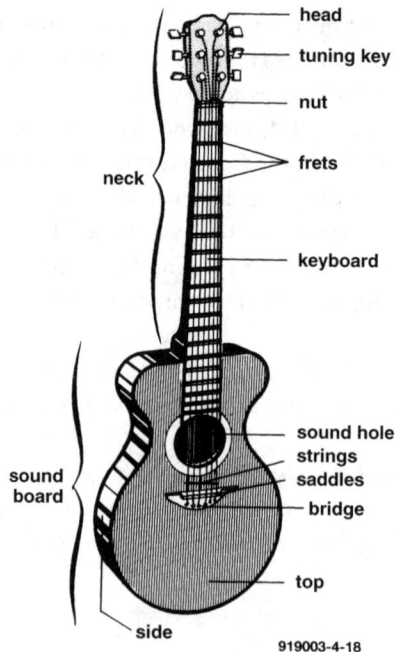

Figure 2.2.1. The various parts constituting a guitar (acoustic or electric)

The mass of a steel high-E string of an electric guitar of diameter 0.965 mm is 6 grams. The length over which the string can vibrate is 0.65 metre (distance between bridge and nut). Provided the string is correctly tuned, its fundamental frequency when it is plucked (not depressed) is 82 Hz. The tension of the string is then

$$T = (0.006/0.65)(2 \times 82 \times 0.65)^2 \approx 104 \text{ N m}^{-2}$$

The tension in the other strings is of the same order. The tension in nylon strings is appreciably lower, typically 80–85 N m^{-2}.

It does happen that a well-tuned guitar does not sound quite right, which may be due to a number of reasons. Often, the strings are old and dirty or not suitable for the guitar. When a guitar is played hard and often, the strings do not last more than a couple of days. It is also amzaing how many guitar players use incorrect strings. For instance, a classical guitar should never use steel strings. This type of guitar cannot withstand the enormous tension of the strings, which results in the keyboard becoming distorted.

Strings that are too thin cut into the nut in due course. This results in a gradual reduction of the space between strings and keyboard which causes spurious sounds. Moreover, the string is held tight in the cut so that it can no longer be tuned correctly. The only remedy then is to replace the nut. The frets may also be at risk. In some guitars, the frets are made of fairly soft material which causes them to wear out in a releatively short time if steel instead of nylon strings are used.

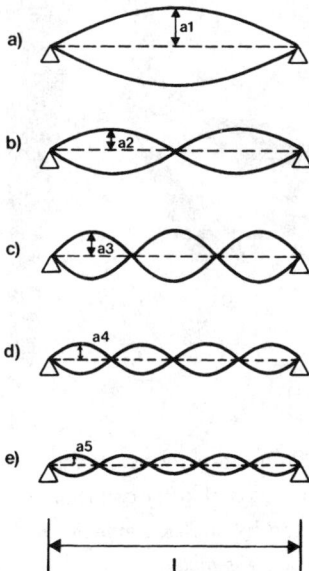

Figure 2.2.2. When a string of a guitar is plucked, not only a fundamental frequency, but also a number of overtones (harmonics) ensue. The fundamental frequency f_1 with amplitude a_1 is shown in a). The wavelength, $\lambda = 2l$. The overtones, which occur simultaneously, are shown in b) to e). Note that the 1st harmonic is the 2nd overtone, the 2nd harmonic, the 3rd overtone, and so on.
b) 1st harmonic: $f_1 = 2f_0$; $\lambda = l$
c) 2nd harmonic: $f_2 = 3f_0$; $\lambda = 2/3l$
d) 3rd harmonic: $f_3 = 4f_0$; $\lambda = 1/2l$
e) 4th harmonic: $f_4 = 5f_0$; $\lambda = 2/5l$

Figure 2.2.3. Photograph of a piezo-electric pick-up element from Shadow, which is here mounted on the bridge of a concert guitar. The photograph also shows the socket for a 3.5 mm audio plug.

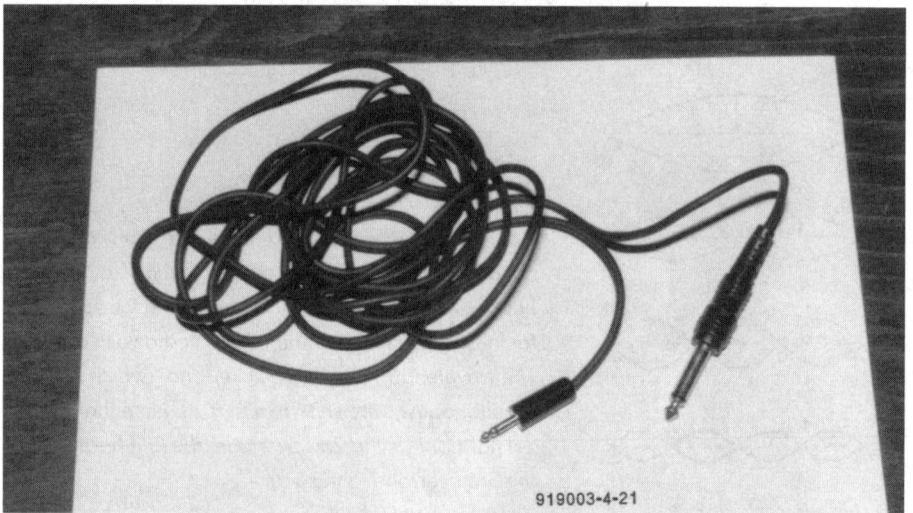

Figure 2.2.4. Connecting cable for the element in Figure 2.2.3. It is terminated at one end into a 3.5 mm audio plug and at the other end into a 6.3 mm audio plug for connection to a standard guitar amplifier. The 6.3 mm plug may be replaced by an XLR connector for connection to a vocal installation or a public-address mixer.

If all is well, a string depressed at the 12th fret should produce a tone at exactly twice the frequency it produces when plucked open. In most acoustic guitars this is not entirely so, but in practice this is normally not a problem because this type of guitar is usually played in the lower ranges.

An acoustic guitar is a particularly soft sounding instrument, which means that when it is used within a group its output must be magnified. The simplest way of doing so is to pick up its sound with a microphone whose output is linked to an amplifier. The fundamental frequencies of an acoustic guitar range from 82 Hz (high-E string) to around 659 Hz (E_4 string at 12th fret). To this must be added the overtones, of course. It is clear from this that an acoustic guitar can be recorded readily with a voice microphone that has a fairly straight frequency response.

Classical guitars with nylon strings produce relatively few overtones, so that the microphone must be aimed at the top of the sound board in the area around the bridge, since that is where most overtones originate.

Western guitars have steel strings and thus produce many overtones. With this type of guitar, the microphone should be pointed at the sound hole. The sound becomes particularly transparent when a condenser microphone is used (for instance, an electret with cardioid polar diagram). If a subminiature microphone is used, it may be clamped to the edge of the sound hole. It is obvious that such a microphone must not be sensitive to proximity effects since this would cause the low frequencies to be overemphasized. In that case, it is better to use an omni-directional microphone, although this type is not ideal as far as any feedback is concerned.

An advantage of recording the guitar via a microphone is that the original sound of the instrument is retained, at least as far as the amplifier permits. There are drawbacks too, however: the guitarist must sit and play directly in front of the microphone, which requires much discipline. Also, the microphone is a potential source of interference because background sounds may also be recorded (crosstalk), and there is a risk of feedback. To remedy these drawbacks, some manufacturers have designed special guitar pickups. Of the many that have been on the market over the years, only those that are fitted between the bridge and the top of the box are still available. They are all based on the piezo-electric effect. Unfortunately, the output of these pickups no longer resembles the original sound of the instrument. However, they are very reliable and give the guitar player much more freedom to move about.

Some types of pickup are provided with a 3.5 mm socket (Figure 2.2.3), whereas others require such a socket to be fitted somewhere in the body. It is advisable to study catalogues of various manufacturers, as well as relevant literature, to decide on the correct type for a certain application. In many cases, fitting the socket is not too difficult, but its location is often dictated by the type of pickup.

Most microphones are supplied with a connecting cable as shown in Figure 2.2.4. This may be fitted as standard with a 3.5 mm plug and a 6.3 mm plug for connection to a standard guitar amplifier, but there are cables with an XLR plug instead of a 6.3 mm plug for connection to a public-address mixer.

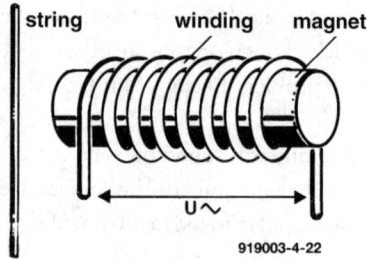

string   winding   magnet

U∿

919003-4-22

Figure 2.2.5. Principle of operation of a pick-up for an electric guitar.
The magnet produces a static magnetic field which is disturbed by the vibrating string.
The consequent variations in the fieldstrength induce
a low-frequency alternating voltage in the coil (see also Section 1.1.1).

There are pickups that need to be fitted against the sound hole of the guitar; these work only with steel strings. The sound produced by these pickups is often quite terrible and is more reminiscent of a bad electric guitar than an acoustic guitar. This is not surprising since the design of this type of pickup is identical to that for an electric guitar.

Ovation guitars are different. These are acoustic guitars in which a piezo-electric element is built into the bridge. The body of the guitar is not made of wood but of lyrachord which suppresses any tendency of feedback. Lyrachord was developed for aircraft construction. There are other makes of guitar available in which this material has been used for the body.

The only similarity between an electric guitar and an acoustic guitar is in the position of the

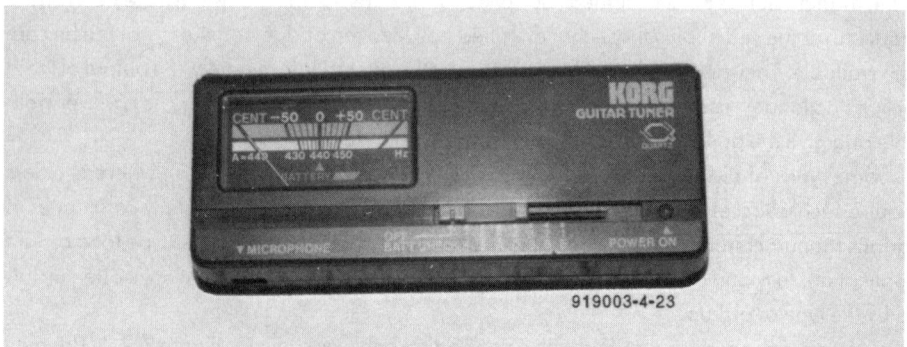

Figure 2.2.6. Electronic tuning unit from Borg. This is normally used with the pointer at the centre of the scale (note A, f = 440 Hz). The unit has an integral microphone which enables, at least in quiet surroundings, an acoustic guitar to be tuned.

Figure 2.2.7. Making the A audible with a tuning fork. The vibrating fork is held close to the pick-up, whereupon the tone is clearly audible via the guitar amplifier.

tones on the keyboard. Otherwise they are two completely different instruments. The frequency range of an electric guitar extends further upwards than that of an acoustic guitar: the highest overtone is 1174.1 Hz. Since an electric guitar has no sound board it hardly produces any sound. The vibrations of the strings are converted into electrical voltages with the aid of an inductive pickup, the principle of which is shown in Figure 2.2.5.

A permanent magnet produces a static magnetic field. Around such a magnet a small coil has been wound. Because of the vibrating of the string the magnetic field changes slightly. This induces a small alternating potential in the coil. This voltage is larger when (a) the string vibrates more strongly, (b) the magnetic field is stronger, (c) the coil has more turns, (d) the distance between the string and the magnetic pole is smaller. This distance must not be too small, however, since then the vibration of the string may be affected by the magnet. In the electric guitar the pickup has special significance since, in contrast to an acoustic guitar where it is a necessary evil, it is responsible to a very large degree for the quality of the sound. Before this pickup is discussed in more detail, we must have a look at the tuning of a guitar.

Figure 2.2.8. Bridge of an electric guitar. The saddles under each string may be adjusted independently. The knurled screws at the sides of the strings enable the distance between strings and keyboard to be set correctly.

An electric guitar may be tuned with the aid of a traditional tuning fork or an electronic tuning unit as shown in Figure 2.2.6. The tuning fork should be held close to the pickup whereupon the tone produced by it is made clearly audible by the amplifier—see Figure 2.2.7. It is, however, better if the reference tone is produced by a wind instrument, provided that the band uses wind instruments. In noisy surroundings the guitar can be tuned only with an electronic tuning unit. In contrast to an acoustic guitar, the tone intervals in an electric guitar can be set quite precisely. Each string has its own adjustable saddle—see Figure 2.2.8. These must be adjusted in such a manner that the twelve frets (twelve semitone intervals) correspond exactly to an octave (doubling of frequency). This may be carried out with the tuning unit. Most of these units produce fundamental frequencies, which means that both the tone of an open plucked string and that of the string depressed at the 12th fret are indicated. When the latter is too low,

919003-5-1

Figure 2.2.9. This photo shows how the length of the string is set with the aid of an electronic tuning unit. First, the string is loosely plucked and tuned in the usual manner. Next, it is plucked at the 12th fret. In both cases, the pointer of the tuning unit must be at the centre of the scale. If it does not reach the centre, the distance between the 12th fret and the bridge is too large. To reduce it, the saddle under the relevant string is moved into the direction of the pickup with a screwdriver. If the pointer overshoots the centre of the scale, the distance has to be increased. This procedure of tuning, checking the tone interval, and setting the saddle, must be repeated as often as necessary until the pointer is at the centre of the scale at all adjustments. These adjustments must be carried out for each and every string. Every time the strings are replaced, their length must be checked and corrected if necessary.

the saddle under the relevant string must be moved closer to the key as shown in Figure 2.2.9. Every time the strings are replaced, the tone interval must be checked and corrected if needed. It is amazing how many guitarists find this a tedious job. Many of them do not even notice when the tone interval is incorrect or know how to correct it when they do notice. Apparently, most musicians are not technically minded or interested. It is up to the sound technician to carry out all these little jobs or at least to tell the musician how they are done.

Apart from the tone interval, the distance between the strings and the keyboard is impor-

Figure 2.2.10. Bridge and pickup on an electric guitar from Aria. Since these form an entity, adjusting the saddles is slightly more complicated.

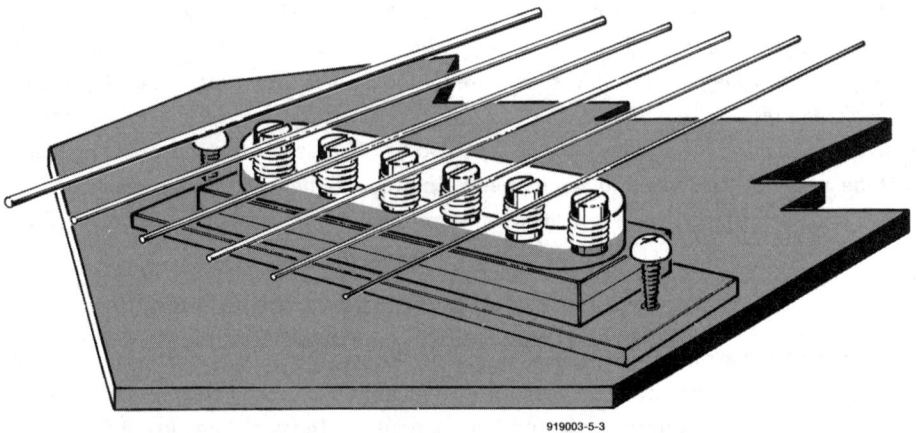

Figure 2.2.11. Guitar pickups can be tilted with the fixing screws. This enables differences in sound level between strings to be compensated.

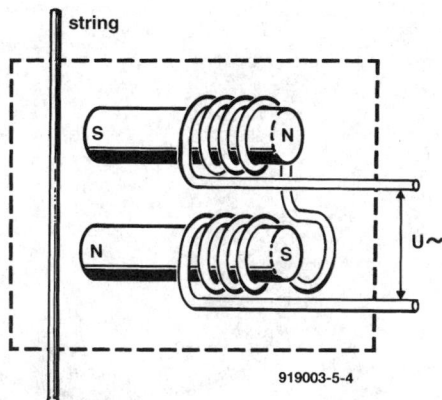

*Figure 2.2.12. Principle of the humbucking (hum-suppressing) pickup. In this,
the coils are connected in series and anti-phase, so that voltages induced
by external magnetic fields nullify each other. Since the the coils have opposite polarity,
the voltages induced by the vibrating strings in them are added together.*

tant. When this is too large, it is difficult to play the guitar properly; when it is too small, spurious sounds may ensue. As so often, a compromise has to be found. The distance between strings and keyboard is set with the bridge, normally with the aid of knurled screws, and by the use of a nut of the correct height. A new nut may be ordered (via the supplier of the guitar).

In electric guitars, the pickup influences the quality of the sound greatly: the guitar itself contributes much less to the sound than an acoustic guitar. Basically, there are two types of pickup. One of these provides a small bar magnet under each string. The south poles of this magnet point to the string (although the polarity of these magnets in principle does not matter). The pickup coils are wound around these magnets and are made from enamelled copper wire of not more than 0.06 mm diameter. In total, there are 7,600 turns, whose resistance is about 5.6 kΩ. The coils are held in place with wax. Damage is prevented by a small plastic cover—see Figure 2.2.10.

In some types the six individual bar magnets are replaced by a single flat magnet, into which a screw is inserted for each string. The advantage of this design is that the distance between string and magnet may be set discretely for each string: in models with individual magnets this distance is set by tilting the whole pickup—see Figure 2.2.11.

Both models are called single-coil pickups. They are typified by a shrill sound that is rich in overtones. They have an annoying property also: they are highly susceptible to stray hum from mains power lines. An alternating magnetic field as caused by each and every transformer induces a hum voltage in the coils. When a guitar so fitted is used in the vicitinity of an electronic equipment, for instance, the guitar amplifier, a disagreeable 50 Hz tone is heard.

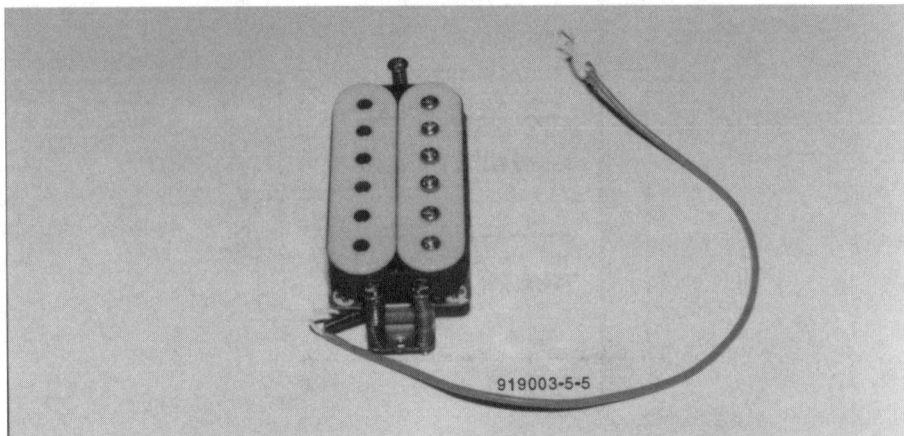

Figure 2.2.13. Humbucking pickup used in a Yamaha Type SG300 electric guitar.

The pickup cannot be fully shielded because it would then not be able to react to the vibrations of the strings. The only, and simplest, solution to this problem is to stay well away from electronic equipment.

The other type of pickup is the humbucker, in which two coils are wound in opposite directions and connected in series as shown in Figure 2.2.12. This type of arrangement ensures that voltages induced in the coils by external magnetic fields cancel each other. The humbucking pickup is therefore much less susceptible to hum.

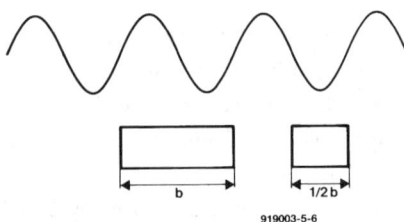

Figure 2.2.14. This illustration shows why a broad pickup collects fewer overtones. In this figure, the large pickup (width b) picks up the whole vibration. Since the peaks and troughs of a wave cancel one another, the pickup does not produce an output voltage. A large part of the vibrations whose wavelength is smaller than the width of the pickup are therefore not picked up. In short: a wide pickup picks up fewer high frequencies. The narrow pickup (width 1/2b) 'sees' only a short length of the string and is therefore able to pick up higher frequencies.

To prevent the voltages induced by the vibrating strings also cancelling one another, the north pole of one coil and the south pole of the other point to the strings. This ensures that the audio-frequency voltages induced in them are added together, so that the output voltage of a humbucking pickup is appreciably larger (theoretically twice as large) than that of a single-coil pickup. A humbucking pickup used in the Yamaha Type SG300 electric guitar is shown in Figure 2.2.13.

Partly owing to their larger dimensions, humbuckers produce a sound that is less rich in overtones than a single-coil pickup (see Figure 2.2.14). The sound is silkier, less sharp. Because the output voltage is fairly high, amplifiers may easily be overdriven. There are also humbucking pickups on the market in which the coils are mounted on top of each other, which gives the pickup the shape and dimensions of a single-coil pickup. Even the sound is reminiscent of that of a single-coil pickup. They are more immune to hum than a single-coil model, however. This type of pickup is marketed by some manufacturers under the name Stack Pickup.

The pickups discussed so far are passive types, that is, they contain no magnifying components. There are also active pickups which amplify and/or alter the sound within the body of the guitar. These pickups normally have a low-impedance output ($100–1000 \, \Omega$). This arrangements enables a great many sound variations. The requisite electronic circuits need to be powered by a batery in the guitar.

A pickup must be designed and constructed with great care to ensure good performance. There are, however, vast differences in quality between the many pickups on the market. If the magnet or coil is not fixed properly, or the turns of the coil are too loose, microphony may ensue. When that happens, the pickup does not only collect the vibrations of the strings, but also noises resulting from holding and playing the guitar. It is also highly susceptible to feedback from loudspeaker to guitar via the loose parts, about which the guitarist cannot do much if anything. Immunity from microphony is therefore an important criterion in judging the quality of a pickup. There are pickups that have an ornamental tin-plate housing, but this normally causes much trouble since it acts as a diaphragm and so engenders feedback. It is almost always better to remove the housing altogether (if particularly wanted by the guitarist, replace it with a DIY suitably varnished plastic or cardboard housing).

Whether a pickup suffers from microphony is established fairly easily: open the volume of the amplifier and tap against the pickup. The softer the tap sounds, the less the pickup suffers from microphony.

In a good-quality guitar, feedback via the strings is much more common than feedback via the pickup itsef. This can be tested by holding the guitar directly in front of the loudspeaker and increasing the volume of the amplifier until feedback occurs. If the strings are at fault, the feedback disappears when they are held; feedback caused by the pickup cannot be so suppressed.

The terms 'hi-fi' and 'stereo' applied to pickups have no meaning whatsoever as far as musical instruments or performing music are concerned: they belong to the field of audio engineering.

It is not only the construction of the pickup, but also its location on the guitar that affects the

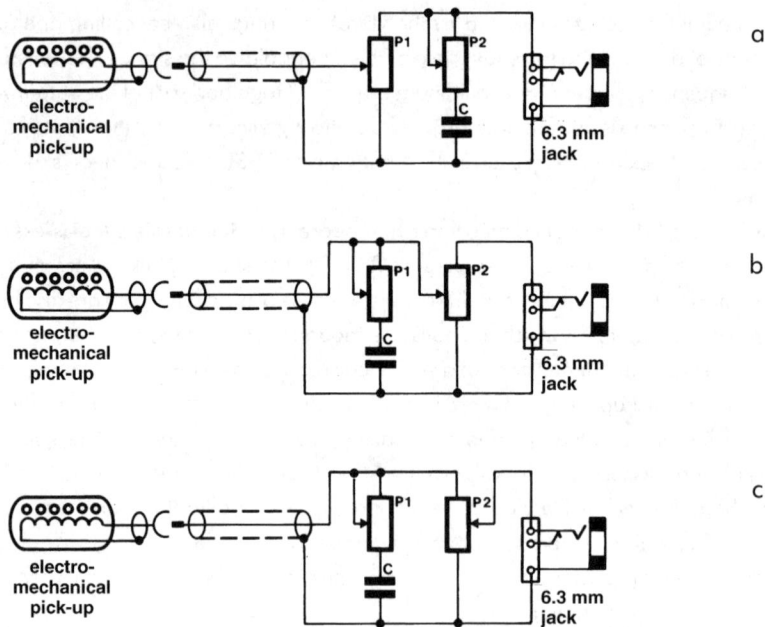

electro-mechanical pick-up

6.3 mm jack

a

electro-mechanical pick-up

6.3 mm jack

b

electro-mechanical pick-up

6.3 mm jack

c

919003-5-7

Figure 2.2.15. Some typical, standard circuits often found in electric guitars.

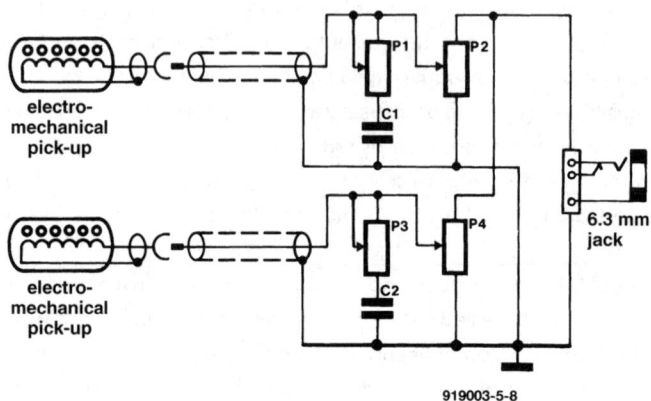

electro-mechanical pick-up

electro-mechanical pick-up

6.3 mm jack

919003-5-8

Figure 2.2.16. Two parallel-connected pickups connected to a common output socket. This circuit is found among others in the Yamaha SG300 electric guitar.

sound. When it is placed near the bridge, the sound is shrill and rich in overtones; further away from the bridge in the direction of the nut, the sound becomes muffled, more closed-in. This is why most electric guitars have two or even three pickups mounted at various positions along the strings that favour various harmonics. The front one of these, that is, the one nearest the fretboard is more sensitive to higher harmonics. This arrangement allows the guitarist to mix the desired signals from the pickups by switches or gain controls.

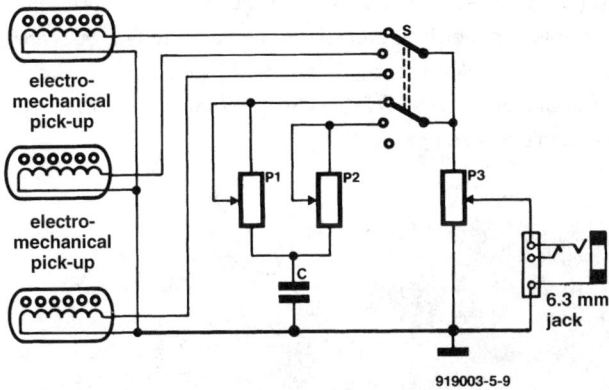

919003-5-9

Figure 2.2.17. Circuit originally used in the Fender Stratocaster electric guitar for linking three pickups to one common output socket.

The sound technician needs to be acquainted with the electronic circuitry within the guitar, becaus this may become defect or need to be modified to individual requirements. Some standard circuits often found in guitars are shown in Figure 2.2.15. In circuit a, the pickup is linked to volume control P_1 via a screened cable. Potentiometer P_2 is a simple tone control. As stated in 1.1.9, a potentiometer has three terminals, of which the two outer ones are connected to a carbon track inside the component, while the centre terminal is the wiper that is moved across the track with the external control knob. When the wiper is turned fully anticlockwise, it is at earth potential and no signal is passed to the remainder of the circuit.

The tone control only allows high frequencies to be attenuated. When the control is turned fully anticlockwise, the wiper short-circuits the carbon track and high frequencies are increasingly shorted to earth by capacitor C, whereupon the sound becomes more muffled.

The circuit in b differs from that in a only in that the positions of the volume and tone controls have been interchanged, while that in c differs from that in b by the different wiring of the wiper (see also 1.1.9). In all three circuits, only one pickup can be linked to the output socket. Figure 2.2.16 shows a configuration in which two parallel-connected pickups are linked to a common output bus. It has a drawback in that the two tone controls interact when the volume levels are fairly high.

Figure 2.2.17 shows the circuit originally used in the Fender Stratocaster electric guitar. Three-position switch S allows one of the three pickups to be linked into the circuit. The upper and centre pickups each have a tone control, the lower one (nearest the bridge) has not. The pickups in virtually all passive guitars are configured in a similar manner.

In active electric guitars, the circuit is rather more complex. The signal from the pickup is amplified and modified in various ways. Whether this is an advantage depends on individual requirements. Manufacturers endeavour to market guitars with active electronics, claiming that the reproduction quality is better. However, it is the sound ultimately radiated by the loudspeaker that is of importance. It should be borne in mind that the electronic circuits within the guitar are subject to the limitations discussed in Section 1.4. Moreover, active electronic circuits need a power supply—in this case, a battery.

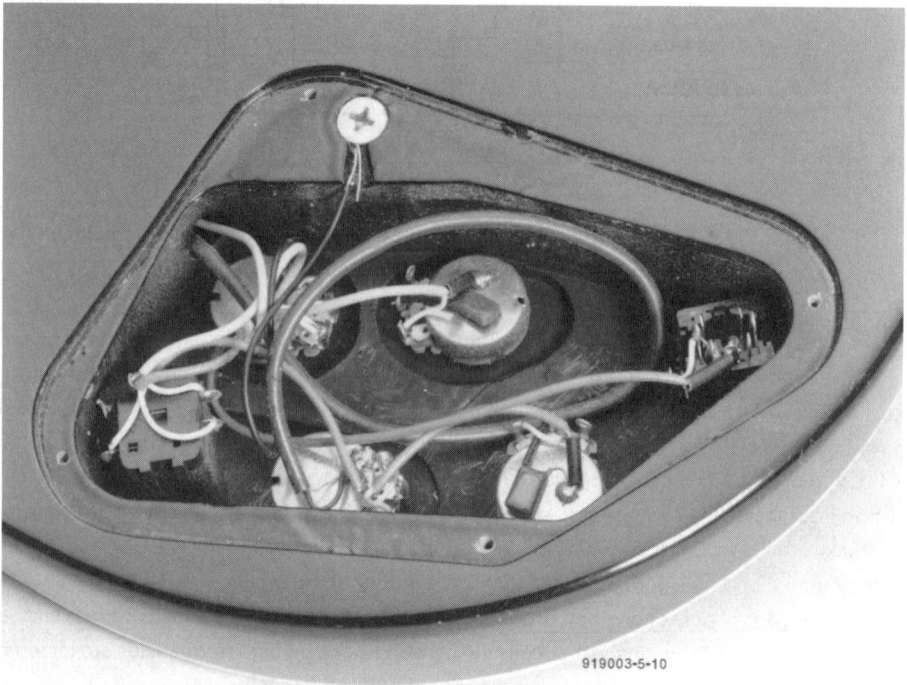

919003-5-10

Figure 2.2.18. In the Yamaha SG300, the electronic circuits are accessible via a removable cover in the backplate of the guitar. The tiny rectangular items on the potentiometers are capacitors C_1 and C_2 from Figure 2.2.16. At the top left is the opening via which the interconnecting cables are passed to the pickups.

The electronic circuits are invariably arranged in recesses in the body of the guitar. That in the Fender Stratocaster is rather larger than in most other models. This is because it houses not only the pickup but also the operating controls. The assembly has a plastic cover into which the potentiometers and switches are fitted. If any work has to be carried out on the assembly, the strings have to be removed from the guitar. In other models, such as the Yamaha SG300, the recess for the potentiometers is just large enough, while the pickups are housed in another recess. The electronic circuits are accessible from the back plate of the guitar—see Figure 2.2.18.

Holes are drilled in the body of the guitar to accommodate the wires to the pickups. In hybrid guitars the electronic circuits are mounted in the body; they are accessible either via the f-holes or via the fixing hole of the pickup (after this has been removed).

Guitars are almost always linked to the amplifier by cables terminated into 6.3 mm audio plugs, which is possible since the cables are relatively short.

When an electric guitar is to be amplified by a public-address system, the use of a microphone is unavoidable. For this purpose, dynamic, mid- or low-tuned microphones should be used. When the microphone location is being determined, the polar diagram of the loudspeaker(s) must be taken into account (see Section 1.3.3). The microphone must not be directed at the centre of the loudspeaker since this would overemphasize the high frequencies. To ensure that the sound collected by the microphone is identical (or nearly so) to that in the rehearsal room, the microphone must be pointed at the edge of the diaphragm—see Figure 2.2.19.

Figure 2.2.19. Position of the microphone for recording the output of a guitar amplifier.
To ensure a 'natural' sound, the microphone must not be pointed
at the centre of the loudspeaker diaphragm.

Figure 2.2.20. Some bass guitars, such as the Jay-Dee model shown here, have two outputs: a balanced XLR plug to facilitate direct connection to a mixer, and a 6.3 mm audio plug for connection to a bass guitar amplifier.

2.2.2 Bass guitar (electric bass)

An important type of electric guitar is the bass guitar, widely used in rock and jazz bands. It generally has four strings tuned in fourths, that is E_1, A_1, D_2 and G_2 (corresponding to frequencies of 41.2, 55, 73.4, and 98 Hz). It also has a longer fretboard (90 cm) than the normal electric guitar.

An electric bass must be recorded via a microphone, just like an acoustic guitar, or with the aid of a piezo-electric pickup. There are special types of this pickup on the market, which perform well with this kind of guitar. Normally, the instrument need not be modified, since the pickups can normally be accommodated to the left and right of the saddles. Such pickups can also be used with other instruments, such as cellos. The choice of a suitable microphone is rather more difficult than with a normal electric guitar owing to the lower frequency range. Acoustic

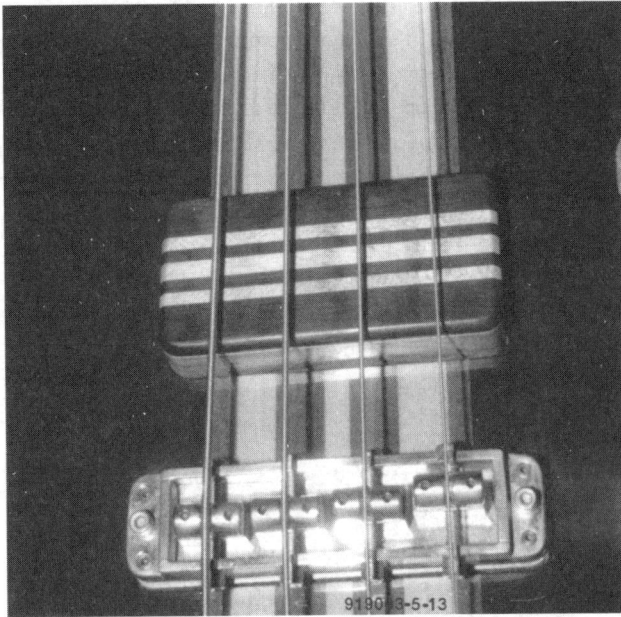

Figure 2.2.21. The bridge and pickup of an electric bass guitar have the same function as in a normal electric guitar.

bass guitars, such as the double bass, are used only in classical music and jazz.

Nowadays, only electric basses are used in rock and pop music. Their operation is virtually identical to that of normal electric guitars, but the frequency range is lower. As in the smaller electric guitar, there are plenty of overtones that, with a certain kind of playing, reach well into the frequency range of the standard guitar. The lower fundamental frequencies result from longer and thicker strings (see Eq. 30). The cable connecting a bass guitar to a mixer is usually terminated into an XLR plug, and that for connection to an amplifier normally into a 6.3 mm jack—see Figure 2.2.20. The pickup and bridge of a Jay-Dee bass guitar are shown in Figure 2.2.21.

The tension, T, of the strings of a bass guitar is appreciably greater than that of a standard electric guitar with steel strings. The length, l, of the strings of a bass guitar is 90 cm, the diameter, d, 2.3 mm, which gives a mass, m, of 25 grams. Provided the string is well tuned, the frequency, f_o, of the plucked (not depressed) string is 41.2 Hz. The tension is calculated with the aid of Eq. 31:

$$T = (0.025/0.9)(41.2 \times 2 \times 0.86)^2 \approx 153 \text{ N m}^{-2}$$

As would be expected, the string tension is greater than in the standard guitar.

When it is required to amplify a bass guitar via a public-address system, use the line output of the amplifier (see Chapter 3, Section 3.2). Since the effect of the loudspeaker on the sound of a bass guitar is much smaller than in the case of a standard guitar, it is much better to take the signal from the line output of the amplifier: most bass players have no problems with this. If, however, the sound is to be recorded via a microphone, use a low-tuned electrodynamic type with a large diaphragm. Again, when positioning the microphone, take into account the polar diagram of the loudspeaker (see 1.3.3).

A bass guitar with an XLR output can be plugged directly into the mixer; in that case, the bass player does not need a separate amplifier. Some players do not like this, however, since the sound is then set or determined solely by the mixer operator.

919003-5-14

Figure 2.2.22. Typical percussion section – A: large tom-tom; B: large timbales; C: bass drum; D: small timbales; E: snare drum; F: hi-hat; G: large cymbals; H: small cymbals,

2.2.2 Percussion instruments

In drums, sound is generated by a vibrating membrane (the head) stretched across a resonating space (air cavity). A modern percussion section normally consists of a bass drum, which is struck via a pedal, a snare drum, various suspended drums, a standing drum, a high hat (also with pedal), and two or three cymbals.

Modern drums may be divided into three groups: those that consist of a single membrane coupled to an enclosed air cavity, such as kettledrums; those consisting of a single membrane open to the air on both sides, such as tom-toms and congas; and those having two membranes coupled by an enclosed air cavity, such as bass drums and snare drums.

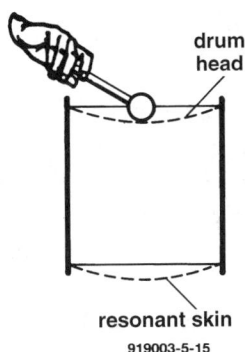

drum head

resonant skin

919003-5-15

Figure 2.2.23. In a correctly tuned drum, the carry or resonating head should move in true synchrony with the drum or batter head

The drum or batter head in top-quality drums is made of calf skin, but in most other drums it is made of Mylar (polyethylene terephtalate). This material, in contrast to calf skin, is not affected by humidity and is much easier to tune owing to its homogeneity. The carry or resonating head is made of similar material. The resonating head of a bass drum has a small hole. A drum with a carry head must be tuned carefully, since a wrongly tensioned carry head leads to disturbing interference. When the drum head moves downwards, the carry head should move downwards in synchrony (see Figure 2.2.23). Accomplishing this requires a great deal of experience. Moreover, many drummers tune the batter or beating head to a greater tension than the resonating head. Although a distinctive timbre results from setting both heads to the same tension, the prominent overtones in the lower range appear to be stronger, at least initially, and to decay more rapidly when the resonating head is tuned below the batter head.

In contrast to other instruments, percussion instruments in general, and the drums in particular, have no fundamental tone. The generated sound is rather more like noise in that it occupies a wide frequency range. The frequency depends on the surface area of the membrane and the dimensions of the body. The width of the frequency range and the loudness of the drums

Figure 2.2.24. This is how the microphone should be placed for recording a bass drum.

919003-5-16

depend largely on the construction of the instrument and the nature of the membrane. For instance, double membranes separated by a thin film of oil sound sound more mellow, sweeter, produce fewer overtones and have a narrower frequency range than single membranes.

Cymbals generate a very wide range of frequencies. In contrast to drums, there is no discernible fundamental tone. The frequencies are generallly higher than those produced by drums: to well above 10 kHz. Nevertheless, low frequencies, down to 800 Hz, are also generated. The sound of cymbals swells only gradually: the maximum sound level is reached only after about 400 ms.

The triangle produces even higher frequencies than cymbals: up to 17 kHz. The instru-

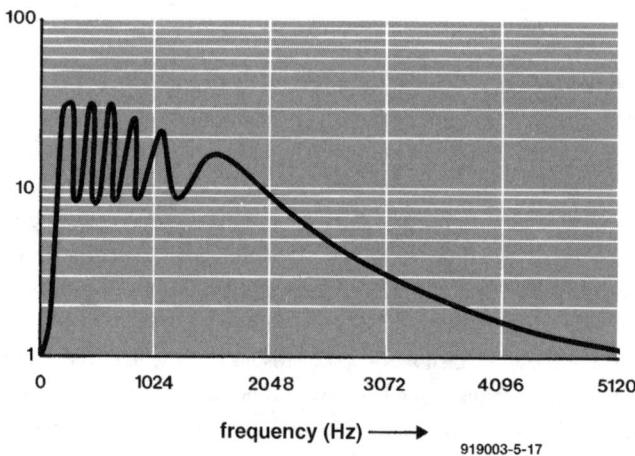

frequency (Hz) ⟶

919003-5-17

Figure 2.2.25. Frequency spectrum of a snare drum: instead of discrete frequencies, a broad, continuous spectrum (up to 5 kHz) is generated. Certain frequencies are produced more strongly than others.

110

ment is, however, not used in pop or rock music.

The percussion section of most pop and rock groups is by far the noisiest, which makes it unnecessary for these instruments to be amplified during performances in small halls or during rehearsal. The fact that sound quality deteriorates through unnecessary amplification is particularly true of percussion instrument (see 1.4). If the dimensions of the hall make it necessary for the percussion instruments to be amplified by a public-address system, use different microphones to record the separate instruments. From an acoustical point of view, the instruments are distinguished primarily by their frequency range and the maximum sound level. The fundamental tone of a bass drum lies, depending on the diameter of the body and the tension of the membrane, between 40 Hz and 60 Hz. Overtones are mainly determined by the material of the membrane and the stick that hits the drum head via the pedal. They normally range up to 200 Hz, to which must be added all sorts of spurious sound: up to 4 kHz. It must be taken into account that the bass drum can produce very high sound levels which the microphone must be able to handle without difficulties. For instance, condenser microphones are totally unsuitable. There are low-tuned, dynamic microphones available that are specially designed for this purpose. The microphone must be placed about 10 cm (4 in) behind the beating head, but should not point to the centre of the membrane – see Figure 2.2.24. To prevent too long an after-sound, most drummers stick strips of felt, foam or even paper tissues on the beating head.

To prevent snare drums producing spurious sounds, they must be stretched tightly, while it may be necessary to stick strips of tape to the resonating head. The microphone is best placed a few centimetres above the edge of the beating head. Because the sound level here is not so great as with a bass drum, a condensor microphone may be used. Often, these microphones are sufficiently sensitive to record the high hat at the same time, provided the individual placings allow this. Special microphones with a figure-of-eight polar diagram are ideally suited for this purpose.

Tom-toms may have one or two heads and, although they are classified as untuned drums, they convey an identifiable pitch. These drums have a diameter varying from 20 cm to 45 cm (8–18 in) and are 20–50 cm (8–20 in) deep. Their fundamental frequency lies between 50 Hz and 200 Hz, depending on the dimensions. Harmonics (overtones) reach to about 1000 Hz. Since spurious sounds (noise) may occur at frequencies up to 6 kHz, depending on the membrane and sticks, it is best if the microphone is placed a few centimetres (about an inch) above the drum head. There is a likelihood of crosstalk from the cymbals and other drums, however, and it is for this reason that the microphone is sometimes placed in the underside of the drum (which can, of course, only be done after the resonating head has been removed). Almost any type of dynamic microphone with a cardioid polar diagram may be used for recording drums.

Cymbals produce a great many overtones: sometimes up to 20 kHz in the case of a hi-hat. The microphone should be placed a few centimetres above the upper cymbal and never at a height opposite the space between the cymbals. Cymbals are well audible without amplification since their sound pressure level is high. In music, high-frequency sounds normally have

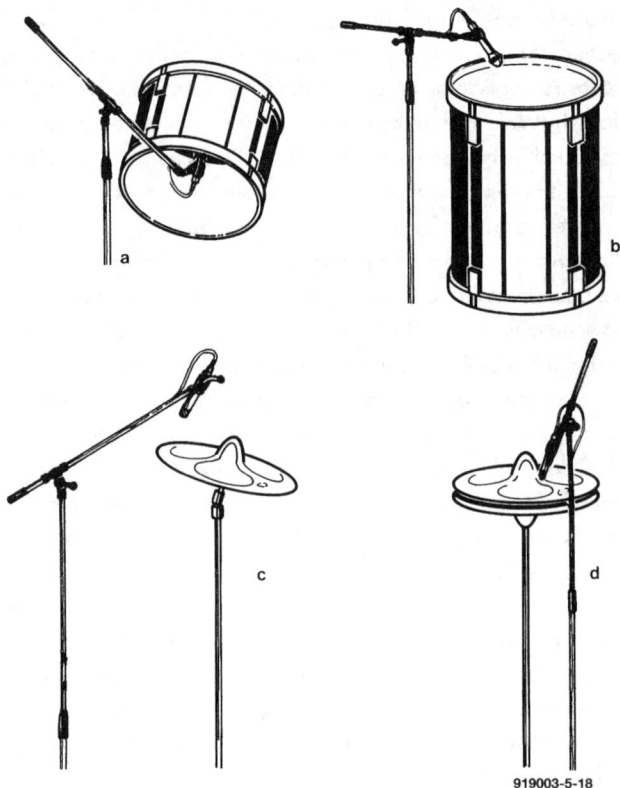

919003-5-18

Figure 2.2.26. Typical microphone positions for amplifying the drum sounds via a public-address system: (a) hanging drum; (b) snare drum; (c) cymbals; (d) hi-hat.

a lower sound pressure level than low frequencies. In most small to medium-sized rooms, cymbals therefore need not be amplified, even if the remainder of the percussion instruments is. However, this is not always so and a check in situ is usually advisable. Correct placing of microphones with various percussion instruments is shown in Figure 2.2.26. Nowadays, there are micro-microphones on the market that can be placed directly on the instruments. Their properties are usually identical to those of standard microphones. They drastically reduce the number of requisite microphone standards and so enable the instruments to be put into position on the stage much more rapidly.

When a public-address (PA) system is used regularly, a small mixer (sometimes called sub-mixer) is very useful for combining the sounds from the various percussion instruments into one or, better, two channels. This is shown diagrammatically in Figure 2.2.27. The correct mixing of the various microphones is one of the more difficult tasks for the mixer operator. Most public-address systems do not reproduce individual drums satisfactorily: only the overall, combined sound is acceptable. Good-quality recordings of percussion instruments can be made only in a sound studio. There, different microphones are used that are placed at some distance from the percussion instruments.

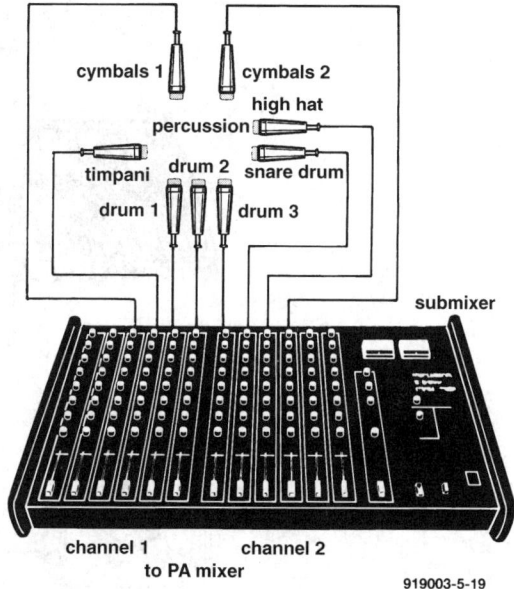

Figure 2.2.27. When a public-address system is used, it is advisable to combine the sounds of the percussion instruments into two channels with the aid of a submixer.

Today, there are also electric percussion instruments on the market. These have no kettle, which, as in the case of an electric guitar, is replaced by an amplifier and loudspeaker. Each beating head is equipped with an electronic pulse generator. The trigger pulses cause preprogrammed tones to be produced. A complete electric drumset consists of the percussion assembly, a mixer with sound generator, an amplifier and a loudspeaker as shown in Figure 2.2.28. The advantages of such a set are that it can be played softly and, at least in principle, any random sound can be synthesized. A drawback is, however, the extensive requisite cabling and the consequent likelihood of interference, the loss of dynamic range, and the loss of the natural, spatial sound: electric percussion instruments sound flat.

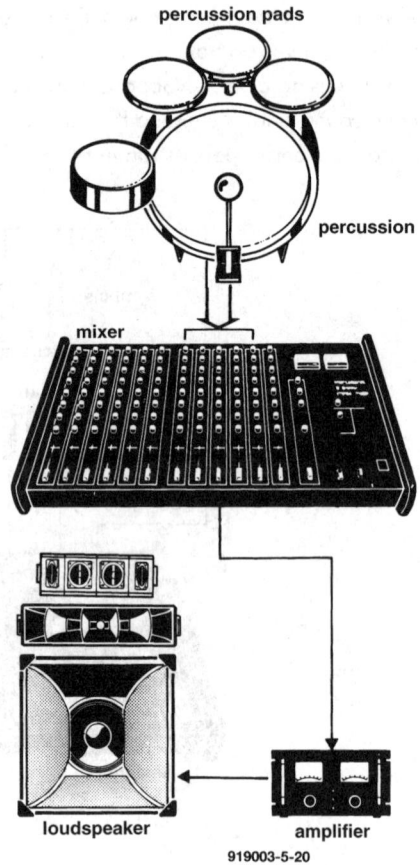

Figure 2.2.28. An electric drumset can produce sounds only with the aid of a mixer with sound generator, amplifier and loudspeaker.

percussion pads

percussion

mixer

loudspeaker

amplifier

919003-5-20

2.2.4 Keyboard instruments

Any instrument that is played with a keyboard is called a keyboard instrument. In principle, however, mechanical keyboard instruments such as the piano are string instruments. In a pipe organ, the sound is produced by large pipes in which a column of air is set into vibration by compressed air. In an electric organ, the sound is produced by means of electronic circuits.

2.2.4.1 Pianos

As mentioned earlier, pianos are pure mechanical instruments. When a key is depressed, an associate string is struck by a small hammer via a transfer mechanism, whereupon the string is set into vibration. Owing to the complex nature of the mechanical construction, any repair or tuning of pianos is better left to qualified personnel.

Although a piano can sound very loud, it may be necessary to amplify its sound. This is done by suspending a microphone in the open piano roughly at the centre of the sound board as shown in Figure 2.2.29. The frequency range of the fundamental frequencies is 27.5 Hz to 4,200 Hz. The overtones reach to well in excess of 10 kHz. It is therefore advisable to use two microphones: a low-tuned one for the low frequencies and a medium-high tuned one for the higher frequencies. The microphones should be pointed at the strings.

Figure 2.2.29. Position of a microphone inside the piano: it is roughly at the centre of the sound board. The use of two microphones gives much better results.

115

2.2.4.2 Keyboard sound generators

The design of keyboard sound generators may be electromechanical, analogue or digital. An electric piano, such as the Fender Rhodes from yesteryear, uses such sound generators to emulate the sound of a standard piano. The vibrating elements may be strings, tuning forks, reeds, and others. Computers and digital electronic circuits have made possible the design of digital electronic pianos, which have virtually replaced electromechanical types.

Digital pianos depend on large libraries of stored piano sounds that are recalled when the relevant keys are pressed. They incorporate key-velocity sensors and simulated hammers to imitate the touch and feel of a standard piano. Also, they include features of electronic syn-

Figure 2.2.30. Electric piano with synthesis sound generation (Technics).

Figure 2.2.31. Old 'string orchestra' (LeLogan).

116

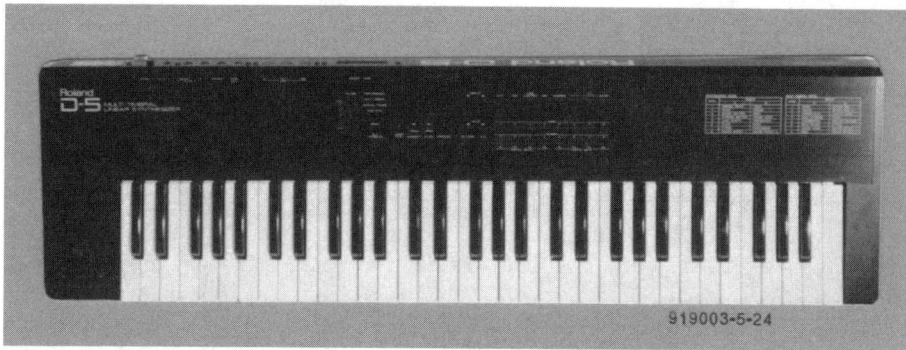

Figure 2.2.32. Modern synthesizer (Roland).

thesizers, such as sequencers, transposers, prerecorded rhythms and accompaniments.

The keys of many keyboard sound generators operate a switch that turns on the appropriate generator. This gives the keys a different feel from those of a standard piano. In some of the more expensive instruments, the keys are therefore 'manipulated" with weights or velocity sensors and simulated hammers to give them the touch and feel of those of a piano.

Apart from the keyboard sound generators designed to produce one specific sound, such as the electric piano or string orchestra, there are others that can produce a number of sounds. Today, the popularity of electric pianos and string orchestras has been replaced by that for synthesizers with preprogrammed and freely accessible sounds. In fact, these are specialized audio computers. Some examples are shown in the photographs in Figures 2.2.30, 2.2.31 and 2.2.32.

It must be said that, basically, the synthesizing of sounds with electronic circuits is fairly straightforward. However, the electronic reproduction of the natural sound of an acoustic instrument is not easy, which explains the high price of synthesizers with a good piano sound. Even then, the dynamic range is much smaller than that of the acoustic piano.

It is unfortunate that the hype word 'stereo' is also used in relation to keyboard sound generators. This makes no sense, since 'stereo' is a concept in transfer technology. Since the sound generator produces sound, and there is no question of transfer, they cannot be 'stereo', but should perhaps be termed 'two-channel instruments'. This would mean that the sound is reproduced via two discrete channels, each with its own amplifier and loudspeaker, which would, of course, make possible interesting spatial effects.

Programmable synthesizers normally make it possible for self-programmed sounds to be stored in a memory. Such a memory normally consists of special magnetic cards, which differ from manufacturer to manufacturer and are not normally compatible. It would have been much better if these manufacturers had used the 3.5 inch diskette used in computers.

From a technical point of view, keyboard sound generators do not present any specific prob-

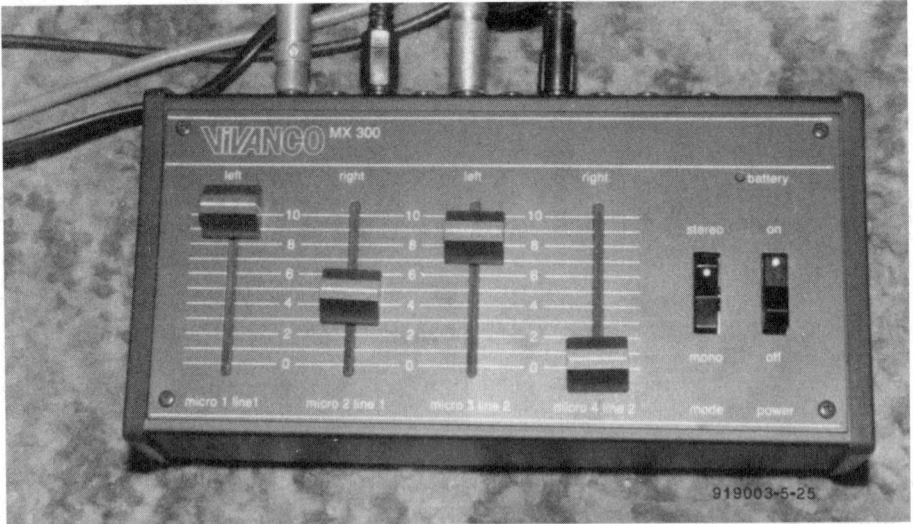

Figure 2.2.33. Submixers combining the outputs of several keyboards and/or other instruments must not affect the sound. Tone and effect controls are superfluous: these would only make the setting up more difficult. The unit shown in this photograph has volume controls only. It can accept two two-channel keyboards, four single-channel keyboards, or one two-channel and two single-channel keyboards. The submixer is powered by a 9-V rechargeable battery, so that it can not impart mains hum to the signal. Also, the absence of a mains cable reduces the number of leads at the back of the keyboard. The current drain is only 3 mA, so that a freshly charged battery has a long life (see 1.1.8).

lems. Their output signal is normally available via a standard 6.3 mm output socket. Most modern keyboards can also be controlled by a computer via a Musical Instrument Digital Interface (MIDI)—see section 10.4. This is, however, primarily of interest for studio recordings.

Like percussion instruments, keyboard sound generators require much attention during the setting up and interconnecting phase, particularly since almost every one incorporates a footswitch for actuating special effects. When several keyboard sound generators are used, it is advisable to combine them in a submixer as shown in Figure 2.2.33 which enables connection to the instrument amplifier via a single cable.

Almost invariably, the signal can be taken from the output of the sound generator or submixer and amplified, if necessary. If rotating loudspeakers (Leslie units) are used, microphones are required. This will lead to difficulties only if the signal levels of the various generators are very different. However, the controls on the submixer can normally equalize such differences in level. In any case, the noise contribution of the mixer must be an absolute minimum so as not to degrade the sound quality.

The practical setup of the keyboards used by a well-known rock group is shown in Figure 2.2.34. The outputs of three keyboard sound generators are combined by a submixer whose output is applied to an amplifier and loudspeaker via a common (foot-operated) volume control. The amplifier is set up just before the performance and requires no further attention. The submixer equalizes the output levels of the sound generators. The synthesizer and

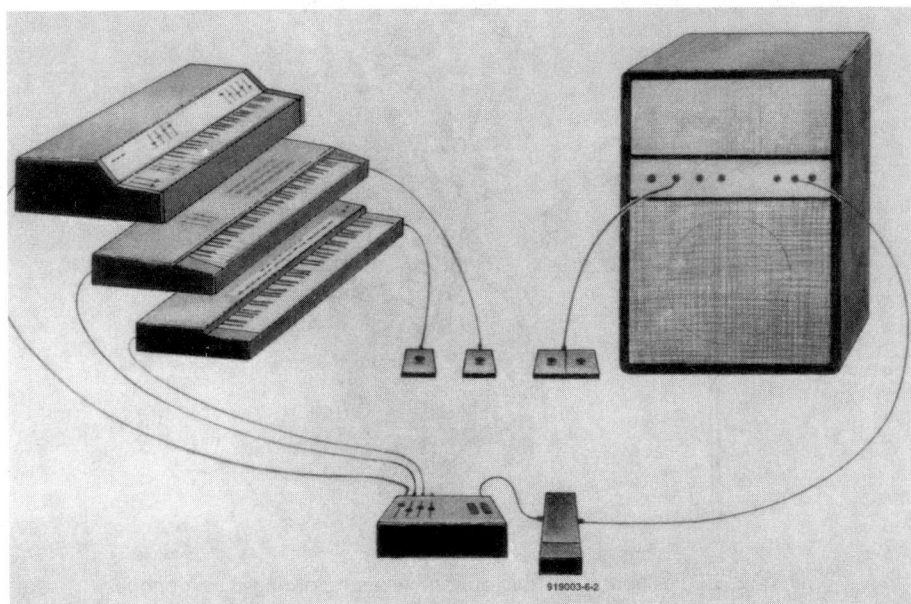

Figure 2.2.34. Interconnecting setup of the equipment used by a well-known rock group.

the electric piano each have an effects pedal. Two further foot-operated switches control the rotational speed of the Leslie loudspeaker(s).

2.2.5 Wind instruments

Wind instruments are purely acoustical and are made to sound by blowing an air jet across an aperture as in whistles and flutes, or by buzzing together the lips or a thin reed and its support, as in trumpets and oboes. Instruments like the trumpet and the trombone produce good sound levels, so that they normally need not be amplified. Others like the mouth organ or clarinet need to be amplified to prevent them being drowned by the percussion section. Usually, voice microphones are perfectly usable, except with wind instruments that produce low notes such as the tuba. Better are miniature microphones that can be clipped onto the instrument.

Many wind instruments have a pronounced directivity, that is, they radiate the sound in one particular direction. This property must be borne in mind when placing recording microphones Note that it is generally not advisable to point the microphone at the centre of the cup of the instrument; generally, a slightly different angle gives much better results—see Figure 2.2.35. The best place is normally found only after a few trials.

919003-6-3

Figure 2.2.35. Position of the microphone when recording a saxophone.

3. Amplifiers

Electric instruments have no sound board: this is replaced by an amplifier and loudspeaker. An amplifier is therefore indispensable for the correct and proper working of the instrument. The singing voice must also be amplified: it cannot compete with the sound level of the trumpet, trombone and percussion instruments. There are special speech amplifiers, which can, however, also be used for amplifying soft acoustic instruments like the classical guitar, clarinet and mouth organ.

3.1 Introduction

Voltage amplifiers for magnifying low-frequency sounds are called audio amplifiers. Their function is, simply, to convert a low-frequency input signal of small amplitude, U_i into an output signal of the same frequency with a larger amplitude, U_o. For the time being, it will be assumed that the shape of the input signal will not be changed, that is, that the amplifier does not introduce any distortion. The relationship between U_i and U_o is constant and, as has been discussed in section 1.2, is called the amplification factor, A. In accordance with Eq. 17:

$$A = U_o/U_i.$$

In voltage amplifiers, the amplification factor is always equal to or greater than 1. Sometimes, the amplification factor is termed (V). Before the technical characteristics of voltage amplifiers are discussed, some other conditions and phenomena which may limit the processing of the electrical signal in general and the amplification in particular must be considered.

3.1.1 Limits of amplification

The magnification of signals in practical, that is, non-theoretical or non-ideal, amplifiers is, as in other branches of engineering, subject to physical limits. It has been emphasized before that any amplifier reduces the dynamic range of the signal. The reason for this is as follows. Each and every active resistance drops a small interference or noise voltage, U_n, which can be calculated with Equation 32.

$$U_n = 2\sqrt{(1.38\times10^{-23}JK^{-1}\times T\times R\times\Delta f)} \qquad [32]$$

121

where T is the temperature of the resistance in kelvin, R is the value of the resistance in Ω, Δf is the bandwidth in Hz, and $1.38\times10^{-23}\times JK^{-1}$ is Boltzmann's constant. The bandwidth is the difference between the lowest and highest frequencies that can be processed. For example, if an amplifier is to be capable of processing signals from 16 Hz to 18 kHz, its bandwidth, that is, Δf, is 17,940 Hz.

Note that the noise voltage, U_n, increases in direct proportion to the bandwidth. This means that a needlessly large bandwidth is a disadvantage since it increases the noise voltage. The noise voltage is also directly proportional to the value of the input resistance. This is why the input resistance in hi-fi equipment is 47 kΩ and in sound engineering not more than 22 kΩ and usually 10 kΩ. The input resistance of microphones is even lower: about 600 Ω.

The noise voltage. U_n, is also called 'white noise' or 'Gaussian noise'. Typical of white noise is that its peak value is always the same, independent of frequency. Apart from white noise, active components in an amplifier, such as valves, transistors, integrated circuits (ICs), and others, also cause a noise signal, which is called 'red noise'. The peak value of red noise, however, is inversely proportional to the frequency (at the rate of 6 dB/octave or 20 dB/decade). Red noise can be reduced appreciably by choosing the lower limit of the frequency range of an amplifier at 60 Hz instead of, say, 10 Hz. The combination of white and red noise is called 'pink noise' (see Figure 3.1.1), which is inversely proportional to frequency at the rate of 3 dB/octave or 10 dB/decade. It is clear that the self-noise of the amplifier limits its input sensitivity, since the amplitude of the signal to be amplified must be (much) larger than the noise. If this is not so, the noise is amplified more than the signal.

There is also an upper limit to amplification. Any amplifier can process only signals whose peak value does not exceed a given maximum. If the maximum is exceeded, even for a fraction of a second, the signal will be distorted. This is called overloading of an amplifier—see Figure 3.1.2.

Virtually all audio recording equipment and some mixers incorporate VU (Visual Unit) meters to enable the operator to keep an eye on the amplitude of the signal. These meters may be pointer instruments or LED (light-emitting diode) bars. The area where overloading occurs is normally indicated in red. It is particularly important in tape recording to adjust the signal level correctly. On the one hand, the signal level must not be too low so as to reduce the influence of pink noise to a minimum and on the other hand it must not be too high – even during signal peaks – to avoid overloading. As so often in engineering, the aim is to find the correct compromise. Simple equipment is normally fitted only with an overload indicator, usually a small signal lamp or LED that lights when a certain signal level is exceeded.

3.1.2 Specifications

Apart from the transfer factor, A, other parameters of amplifiers that are of importance are the frequency, f, versus output voltage, U_o characteristic (at constant U_i); the output impedance,

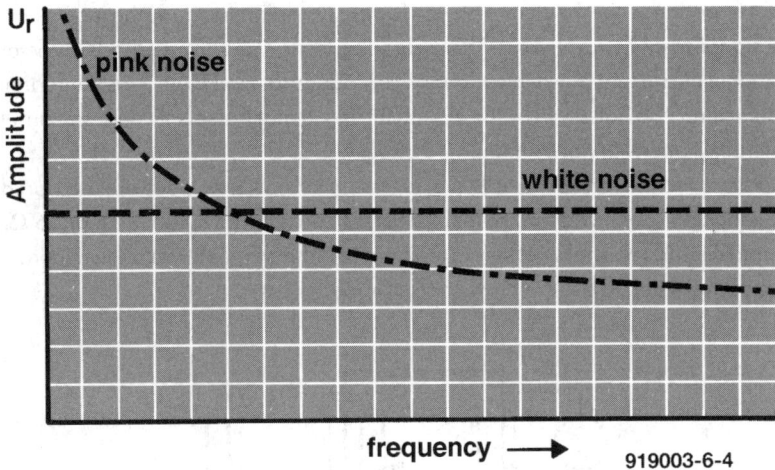

Figure 3.1.1. Noise voltage levels versus frequency characteristics.

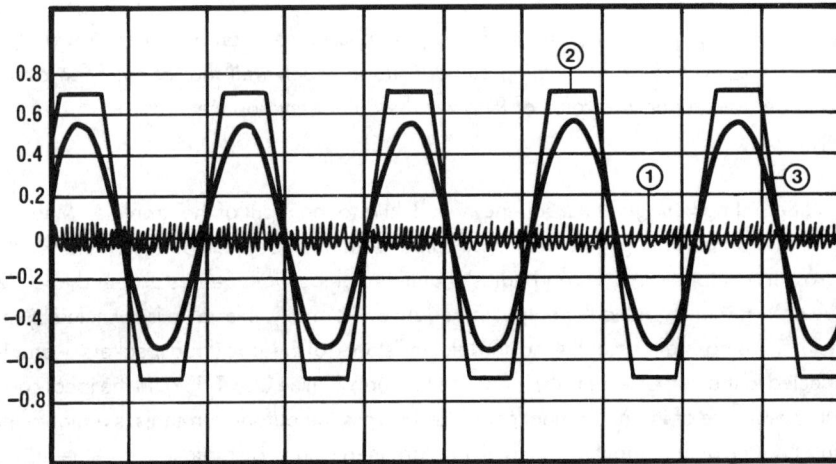

Figure 3.1.2. Limits imposed on the amplification of a signal. Curve 1 shows the self-noise of an amplifier. To ensure undistorted amplification, the level of the input signal must be well above the noise floor. Curve 2 shows what happens when the amplifier is overloaded. The amplitude of the output signal is limited in practice by the peak value of the supply voltage. The amplitude of the signal represented by curve 3 is well clear of the self-noise and the peak amplification, and is therefore amplified without discernible distortion.

Z_o; the input impedance, Z_i; the output power, P_o; the distortion factor, k; the input sensitivity, U_i; and the signal-to-noise (S/N) ratio. The measuring techniques and methods for determining these quantities are laid down in various British Standards and EIA norms. Unfortunately, some manufacturers do not abide by these, with the result that the specifications of amplifiers of different makes often cannot be compared directly.

To interpret numerical values correctly, it is necessary to know how they were measured. The output impedance, Z_o, of power (output) amplifiers is normally $2\ \Omega$, $4\ \Omega$, $8\ \Omega$, or $16\ \Omega$. That of voltage amplifiers lies normally between a few hundred ohms and a few kilohms. Figure 3.1.3. shows a typical test setup to determine the output impedance.

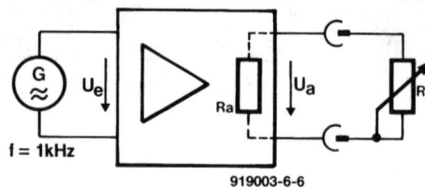

919003-6-6

Figure 3.1.3. Basic circuit for measuring the output impedance of an amplifier. Provided that $R_1 = R_o$, the level of the alternating output voltage is only half that of the open-circuit output voltage. Measuring the value of R_1 gives the output impedance.

Apply an alternating voltage at a frequency, $f = 1$ kHz to the input of the amplifier. Measure the open-circuit output voltage, U_o, across the output terminals. Connect potentiometer R_1 (the load) across the output terminals and adjust it until the output voltage has dropped to $U_o/2$. The value of R_1 is then equal to the output impedance. Normally, this value is specified by the manufacturer. It is normally essential that the total resistance of the load (loudspeakers or amplifier connected to the output terminals) is not smaller than Z_o (see also 1.1.9). In the ideal case, the input impedance of the equipment connected across the output terminals is equal to the output impedance of the amplifier. The two units are then said to be matched, and the amplifier supplies maximum power to the load. If these conditions are not met, the units are said to be mismatched.

Correct matching is of particularl importance in the case of microphones and power amplifiers. In practice, it is desirable for the load (loudspeaker) connected to a power amplifier to be equal to, or slightly larger than, the output impedance of the amplifier. In general, the relationship between the actual output power, P_o, the maximum output power, $P_{o(max)}$, and load impedance, R_l, is

$$P_o/P_{o(max)} = 4Z_oR_l / (Z_o + R_l)^2 \qquad \text{[Eq. 33]}$$

Example: the output impedance of a guitar amplifier, $Z_o = 8\ \Omega$ and the specified maximum output power into an $8\ \Omega$ load is 20 W. If a $4\ \Omega$ loudspeaker is connected to the amplifier, there is a serious risk, particularly in the case of a solid-state amplifier, that it will be overloaded. Yet, according to Eq. 33, the maximum output power is then only

$$P_o = 4Z_oR_l / (Z_o+R_l)^2 \times P_{o(max)} = 4 \times 8 \times 4 /(8+4)^2 \times 20 = 17.8\ \text{W},$$

which is only 89 per cent of the specified maximum power output. If a $12\ \Omega$ loudspeaker were used, the amplifier would not be at risk. The maximum power output is then

$$P_o = 4 \times 8 \times 12/(8+12)^2 \times 20 = 19.2\ \text{W},$$

which is about 96 per cent of the maximum power output.

This example shows that a small mismatch is normally not all that important since the loss in power is small. The loss of 0.8 W in the case of a $12\ \Omega$ loudspeaker cannot be heard. So, do not panic if there is a small mismatch: the consequence of some mismatches are shown in graphical form in Figure 3.1.4.

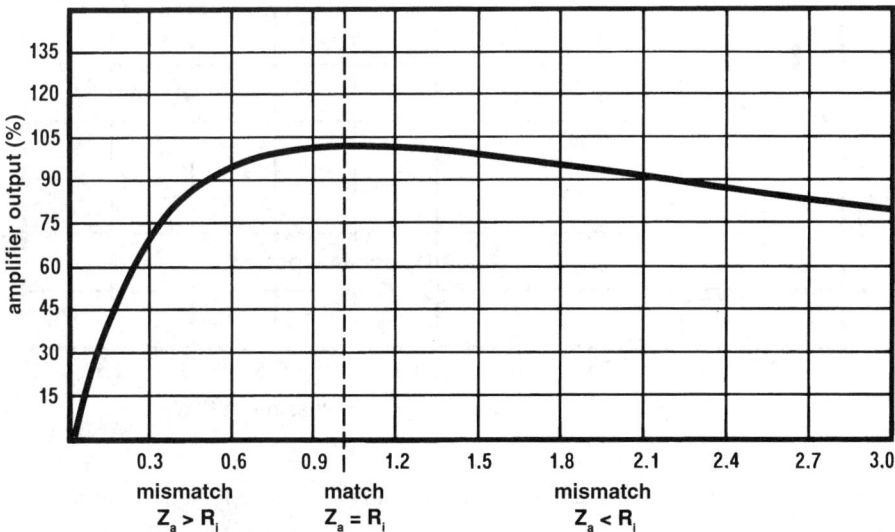

Figure 3.1.4. This characteristic shows that the power loss with an overmatch $(Z_o < R_l)$ is not great. An undermatch $(Z_o\ R_l)$, however, puts the amplifier at risk.

The input impedance, Z_i, is also of importance, since what applies to the matching of loud-speakers to the output terminals applies equally to the matching of a signal source, say, a microphone, and Z_i. The best transfer is obtained with correct matching, that is, when the internal impedance of the microphone is equal to Z_i. This was also seen in Table 2.1.2.(p. 86).

The distortion factor, k, indicates to what degree the input signal is distorted by the amplifier or, in other words, how many harmonics the amplifier has added to the signal (and which were not present in the original signal). This is called non-linear distortion. Measuring the distortion factor is complicated and can be carried out only with fairly expensive test equipment. Normally, the distortion factor is specified by the manufacturers as a percentage. The ideal in hi-fi and public-address amplifiers would be 0 per cent, that is, the amplifier does not introduce any distortion. In practice, this is, of course, regrettably not possible. In the case of music amplifiers, the distortion factor is of interest only in a comparison of the output powers of dif-

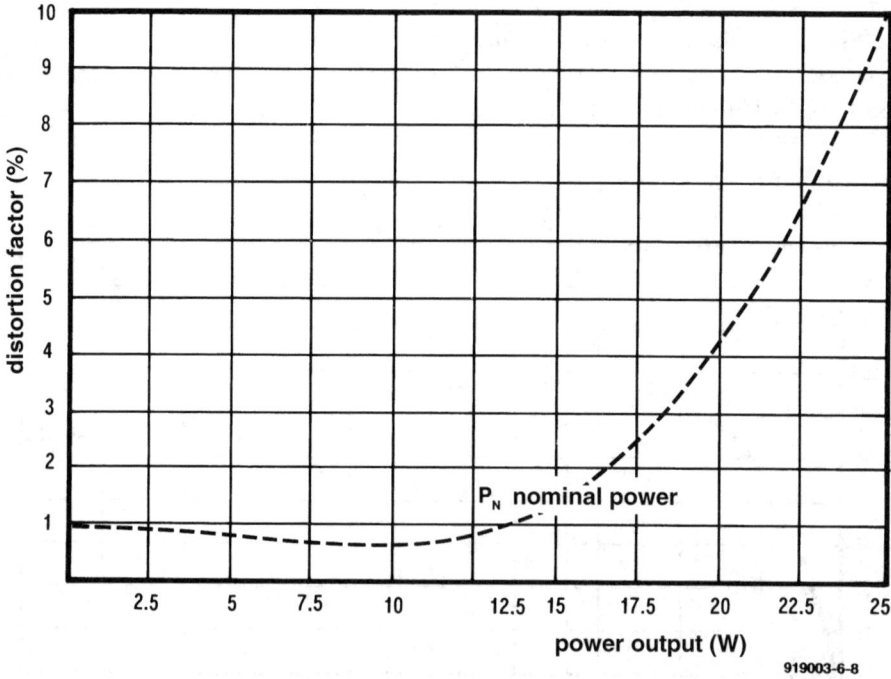

919003-6-8

Figure 3.1.5. Graphic representation of measuring the power output as a function of the distortion factor. The power output is equal to the nominal power output when k reaches a value of 1 per cent; at least, that is how it ought to be. The curve shows that the relevant amplifier has a nominal power output, P_N, of about 13.75 W, assuming the matching is correct—see also Figure 3.1.4.

ferent amplifiers. This is because the nominal power output of an amplifier is closely linked to the distortion factor. This is made clearer in Figure 3.1.5, which shows the relationship between P_o and k.

Normally, the amplifier should supply its nominal power output when $k = 1\%$, but higher values of k are common in musical-instrument amplifiers. However, as stated earlier, the specifications of different amplifiers can be compared only when the various parameters are measured in identical manner. A typical setup for determining the nominal power output is shown in Figure 3.1.6. The generator at the input delivers a signal at a frequency $f = 1$ kHz. The amplifier is terminated into a correct load, which is shunted by a distortion meter. The output of the generator is increased until the distortion factor is 1%. The power is calculated from the values of the load, R_l, and the voltage drop, U_o, across it:

$$P_o = (U_o)^2 / R_l.$$

Note that the output of the generator is a sinusoidal signal and the power output is the RMS (root mean square) value. In the specification of hi-fi equipment, the music power is often stated, but this is not a good criterion and should not be used with musical-instrument amplifiers.

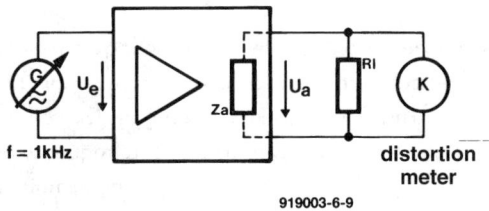

919003-6-9

Figure 3.1.6. Basic circuit for measuring the nominal power output, P_N.
Correct matching ($Z_o = R_1$) is assumed. The input voltage, U_i, is increased until the distortion meter shows a value of 1%. The power output is then calculated from the value of the load and the voltage drop across it.

Another example. It is not strictly necessary to use an expensive signal generator and an even more expensive distortion meter to determine the power output. In this example, a keyboard is used as signal source—see Figure 3.1.7.

The output of the keyboard is connected to the input of the amplifier on test. If possible, a sinusoidal or an organ output is chosen. To avoid excessive sound, the main load is not formed by the loudspeaker but by a load load resistor, R_L. This resistor has the same value and rating as the loudspeaker and is linked across the amplifier output terminals. The loudspeaker, in series with R_2, whose value is at least ×10 the loudspeaker impedance, is in parallel with R_L.

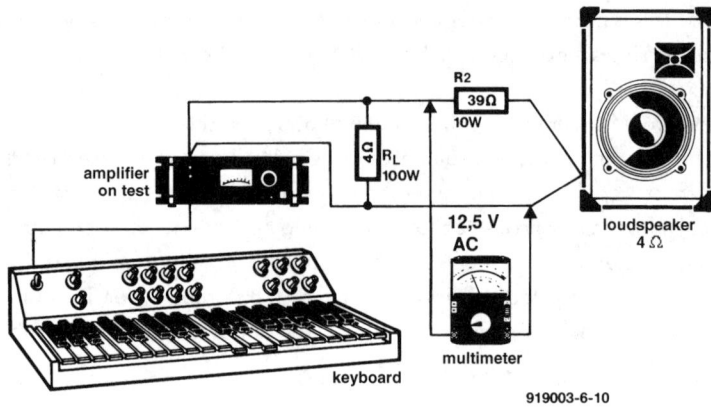

Figure 3.1.7. Practical setup for estimating the power output of an amplifier when a good signal generator and distortion meter are not to hand.

This arrangement ensures that there is only a little sound coming from the loudspeaker. Great care should be taken to ensure that there are no short-circuits anywhere: the amplifier should not be switched on until the setup has been checked thoroughly. The volume is increased until the signal from the keyboard is just not audibly distorted. The frequency of the signal should preferably be near 1 kHz, e.g., B_1 (987.77 Hz). Since many (simple) multimeters cannot measure voltages at that frequency, it may be necessary to use a lower frequency, e.g., A (440 Hz). The multimeter is used to measure the alternating voltage across the load resistor. Since the value of R_2 is large compared with that of R_L, its influence on the measurement may be ignored. In the example, a voltage of 12.5 V a.c. is measured, so that the power output is

$$P_N = P_o = (12.5)^2 / 4 = 39 \text{ W}.$$

It should be noted that such an estimated value is normally on the high side, since 'audible' distortion is highly subjective: most people begin to notice it only when it is higher than 1% (average is 3%).

Load resistor R_L is composed of ten 39 Ω, 10 W resistors in parallel as in Figure 3.1.8. The total resistance is 3.9 Ω (Eq. 15). The total power than can be dissipated is 10×10 = 100 W. Such a composite resistor assembly can be used with most amplifiers, except high-power PA types. It is also very handy when tests or repairs are carried out. Another assembly that is very useful is one of ten parallel-connected resistors of 82 Ω, 10 W each, which gives a total value

128

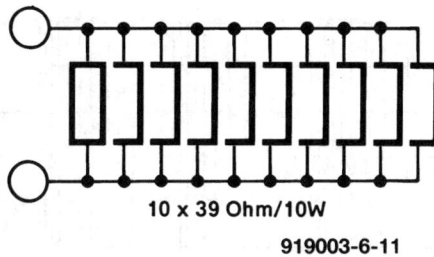

10 x 39 Ohm/10W

919003-6-11

*Figure 3.1.8. Composite 100 W load resistor for assessing the power output of amplifiers.
It is made of ten identical 10 Ω, 10 W power resistors. If the resistors
are vitreous-enamel types, power outputs of up to 150 W can be handled.*

of 8 Ω, which is the terminating load resistance of many amplifiers.

The frequency range or bandwidth states the lowest and highest frequency signals that can be processed by the amplifier without noticeably differing from the nominal transfer level. In hi-fi amplifiers, 'noticeably differing' means not more than ±1.5 dB with respect to the level at 1 kHz (see Figure 3.1.9). Ideally, all frequencies should be amplified equally. Variations in the transfer level as a function of frequency cause linear distortion. In sound engineering, the important power bandwidth is often stated. This gives the frequency range over which at a certain distortion factor (normally 1%) the power output does not drop by more than 3 dB (see Figure 3.1.10). At that power, the voltage across the loudspeaker or load resistance drops to

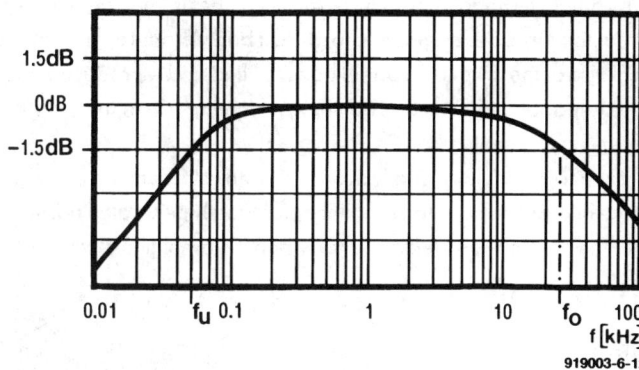

919003-6-12

*Figure 3.1.9. Characteristic to clarify the meaning of frequency range or bandwidth.
Since the curve shows that the lower cut-off frequency, f_1 = 50 Hz and the
upper cut-off frequency, f_2 is 25 kHz, the frequency range is 50–25 000 Hz.*

129

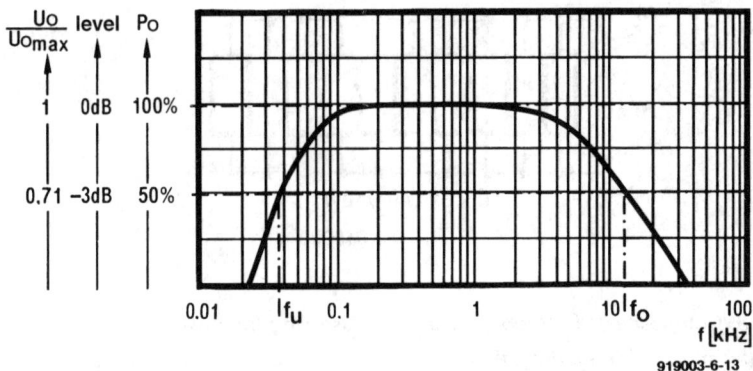

Figure 3.1.10. Characteristic to clarify the meaning of power bandwidth.
In the figure, the power bandwidth is 38–12 000 Hz.

71 per cent of the unloaded level.

In the case of musical-instrument amplifiers, frequency range and power bandwidth are of no importance. They have some significance in voice amplifiers, particularly when these are also used with instruments, for instance, percussion, but are very important in the case of hi-fi and public-address amplifiers since these must reproduce the sound as faithfully as possible.

A useful quantity is the input sensitivity, which is normally specified in terms of input voltage in association with input impedance. The nominal input voltage is the signal level needed at the input to obtain the nominal power output from the amplifier (at 1 kHz, with all volume controls fully open and the output terminals correctly terminated, that is, matched). When the input level is much exceeded, the amplifier begins to clip (distort badly). Input signals at a level close to that of the internal noise of the amplifier cannot be amplified and reproduced properly. This shows once again the importance of correct matching of the input with the output of the signal source (microphone, instrument, recorder, and so on).

The signal-to-noise ratio, which is normally specified in dB, is a measure of the relationship between the level of the internal noise of the amplifier plus locally generated noise and the signal level, for instance, the output of an electric guitar, when the amplifier is driven at full power. The relationship is:

$$\text{S/N ratio} = 20\log_{10}U_o/U_n \quad [\text{dB}] \hspace{3cm} [\text{Eq. 34}]$$

where U_o is the signal level at full power and U_n is the total noise voltage. The larger the S/N ratio of an amplifier, the lower its internal noise.

Example. Manufacturers of hi-fi and PA amplifiers specify as high a value for the S/N ratio

130

as possible. Often, this is measured in circumstances that have little to do with reality. To get an impression of real values, let us consider an amplifier with the usual 47 kΩ output impedance, a frequency range of 20 kHz, and a temperature inside the amplifier of 300 K. The internal noise voltage, according to Eq.32 is

$$U_n = 2\sqrt{(1.38 \times 10^{-23} JK^{-1} \times 300 \times 47 \times 10^3 \times 20 \times 10^3)} = 3.95 \ \mu V.$$

Since most signal sources (microphones, electric guitars, pickups of record players) provide a signal at a level of about 10 mV, then, the signal-to-noise ratio (assuming that there is only internal amplifier noise), according to Eq.34, is

$$S/N = 20\log_{10}[(10 \times 10^{-3})/(3.95 \times 10^{-6})] = 64 \ dB.$$

This is a common and believable value. Amplifiers used in recording studios normally have an input impedance $\leq$ 10 kΩ. The noise voltage is then limited to 1.82 μV, which results in a S/N ratio of 75 dB – a very respectable value.

Musical-instrument amplifiers usually have an input impedance some 2 to 5 times greater than that of hi-fi amplifiers, which gives a noise voltage $\sqrt{2}$ to $\sqrt{5}$ times greater than calculated earlier. This does not necessarily mean that the S/N ratio is bad, since these amplifiers normally have a bandwidth that is much smaller than that of a hi-fi amplifier.

Hi-fi and PA amplifiers are required and intended to reproduce sounds as faithfully as possible. So, such amplifiers may influence the amplitude of the signal, but not the shape of it. In other words, both linear and non-linear distortion must be very low. This situation is quite different in the case of music-instrument amplifiers. These amplifiers are required to influence the shape of the input signal, since the sound finally to be reproduced is, at the input, still in an embryonic state as it were. A musical-instrument amplifier with low linear and non-linear distortion would sound flat and lifeless and would certainly not suffice as the electronic replacement of an acoustic sound board.

3.1.3 Positive feedback

When part of the sound emanating from the loudspeaker is picked up by the microphone and returned to the amplifier, there is a risk of positive feedback, which manifests itself as a whistle. Above a certain sound level, a virtual vicious circle ensues. At every roundtrip, the signal is amplified again until the amplifier produces its peak output—see Figure 3.1.11. The sound installation howls loudly, which can only be remedied by turning the amplifier volume down. Positive feedback can be prevented in one way only: care must be taken to ensure that the sound emanating from the loudspeaker(s) cannot be picked up by the microphone(s). This means that some care must be taken with the location of the loudspeakers and the microphones. For this, it is, of course, essential that the polar diagram of the microphones is known.

Positive feedback does not occur until the sound level reaches a certain value. This level and the frequency of the resulting whistle depend to a very large extent on the dimensions of the room, the furniture, carpets and curtains. The occurrence of positive feedback can be predicted within certain limits, but only with the aid of a very extensive test setup, which is normally only justified in large concert halls and at festivals. There are a number of practical measures which may help to prevent positive feedback.

- The loudspeakers must be placed so that no direct sound can reach the microphone(s).
- If at all possible, there should be no large reflecting surfaces in the direction in which the loudspeakers radiate maximum sound levels.
- Positive feedback at low frequencies is often caused by resonances in the room, combined with the proximity effect. In such a case, it often suffices to increase the distance to the microphone(s) and/or turn back the bass control(s).
- Positive feedback at high frequencies can often be suppressed by turning back the treble control(s). Even when the amplification is still 1–2 dB below the critical level, there is already some indication of the risk of howl: singing voices get a strongly coloured, faint echo. The critical level has then been reached. The relevant tone control should then be turned back to some extent. The frequency at which positive feedback threatens to occur can be suppressed accurately and efficiently by placing an equalizer (Chapter 5) in the signal path.
- Long echoes increase the risk of positive feedback. It is therefore advisable to maintain the values stated in 1.3.4.

Note that positive feedback is not always an unwanted and annoying phenomenon. It may be used in certain circumstances as a modifier: rock guitarists do this regularly. The guitar is then held as closely as possible to the relevant loudspeaker so that the strings are set into vibration by the sound from the loudspeaker so that, once a string has been plucked, a ringing tone ensues (see also 2.2.1).

3.2 Voice amplifiers

A complete voice installation or amplifier consists of one or more microphones and a small mixer, followed by a power amplifier and loudspeaker. The mixer is needed if several microphones are used simultaneously.

3.2.1 General

A concert voice amplifier and its associated loudspeaker are normally designed specifically for amplification of the human voice as far as connections, controls and power out are concerned. It cannot really be used for a different signal source. For instance, if it is connected to a tape recorder, the treble and bass on the tape will be reproduced in an unsatisfactory manner. The overall performance will be flat, colourless, with hiss and overemphasized middle frequencies. There are various reasons for this, one of which is a possible mismatch. The input impedance of a voice amplifier is in the low to medium range, whereas the output impedance of a tape recorder is in the medium to high range. It may also be that the output voltage of the recorder is too high so that the input of the amplifier is overloaded. Also, the frequency response curve of concert voice amplifiers is not straight: it shows a definite rise at the middle frequencies. This results in attenuation of the low and high frequencies, which do not occur in the human voice. The advantage of this is, of course, a smaller likelihood of positive feedback and crosstalk from instruments behind the vocalist.

Voice amplifiers used for pop music normally have a linear transfer characteristic and are therefore suitable for use with a variety of signal sources. This is just as well, because voice amplifiers are used more and more often for amplification of percussion (and also other) instruments. This is because the frequency response can be adapted with individual tone controls to the voice or instrument as the case may be. It could therefore be said that the voice installation nowadays is not so much part of the voice as of the overall instrumentation. Since however singing without amplification in rock and pop music performances is quite impossible, voice

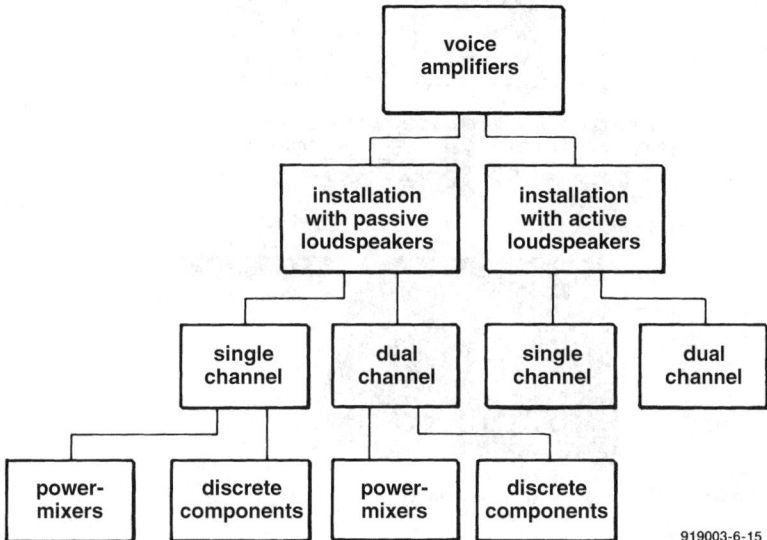

Figure 3.2.1. Classification of voice amplifiers.

amplification in this book is treated as such. Figure 3.2.1 shows some of the many voice installations on the market. Distinction is made between installations with passive loudspeakers and those with active loudspeakers (which usually incorporate the output amplifier).

3.2.2 Installations with passive loudspeakers

Standard voice amplifiers use passive loudspeakers. The complete installation consists, say, of one or more microphones, a mixer, some effects equipment (reverb and echo), sometimes an equalizer, a power amplifier and one or more loudspeakers (see Figure 3.2.2). Strictly speaking, the installation in this illustration is a small public-address system. The transfer to a real PA installation occurs gradually as the mixer gets more facilities and the power output increases. The advantage of such an assembly is that discrete units are easily added or replaced. A disadvantage is the extensive cabling for interconnecting the units.

The cabling becomes much simpler when the installation is constructed as a power mixer

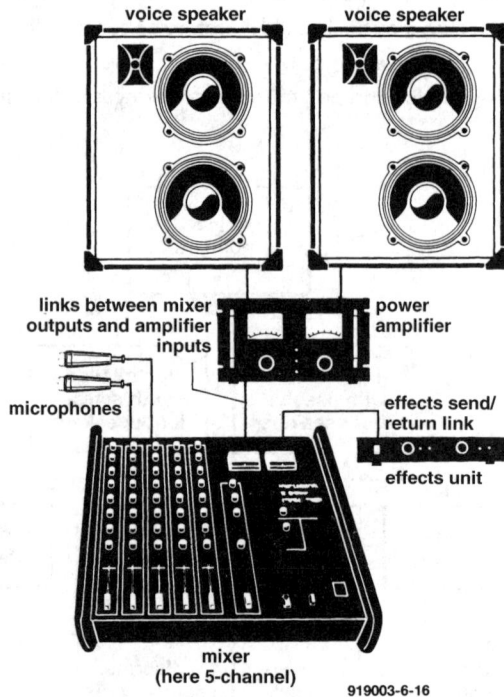

Figure 3.2.2. Voice installation composed of discrete units: microphones, mixer, effects unit, power amplifier and loudspeakers.

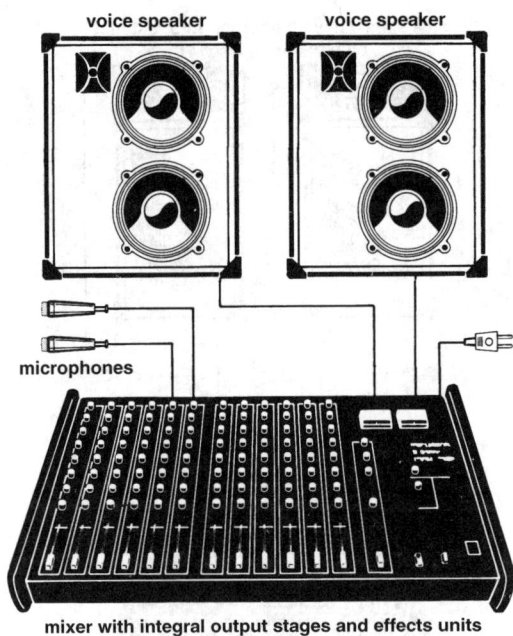

voice speaker   voice speaker

microphones

mixer with integral output stages and effects units

919003-6-17

Figure 3.2.3. In a power mixer all necessary units, such as the mixer, power amplifier and effects unit(s) are housed in a common enclosure.

as shown in Figure 3.2.3. In such an installation, only the microphones, loudspeakers and mains supply need to be connected; the remainder of the equipment is incorporated in the mixer enclosure. Power mixers are near-ideal when regular performances are given in different locations. It should be borne in mind, however, that power output, effects, number of channels, and some other facilities, are fixed.

3.2.3 Installations with active loudspeakers

In installations with active loudspeakers, the output amplifiers are invariably incorporated in the loudspeaker enclosures (see Figure 3.2.4). This means that the loudspeakers can be driven directly by the mixers. The nominal power output is determined by the loudspeakers. A disadvantage of this kind of system, certainly at live performances, is that each loudspeaker needs to be connected to the mixer and to a mains outlet, which increases the number of cables on the stage. However, the output of the mixer can often drive several active loudspeakers simul-

135

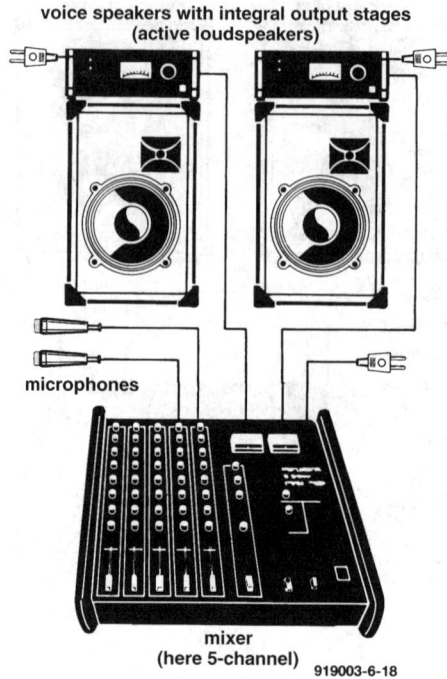

voice speakers with integral output stages
(active loudspeakers)

microphones

mixer
(here 5-channel)

919003-6-18

Figure 3.2.4. Voice installation with active loudspeakers. The loudspeakers, each of which contains an output amplifier, are driven by a standard mixer.

taneously: they are simply connected in parallel. In theory, it is therefore possible to increase the power output of the installation by the use of more and more active loudspeakers. Care must be taken, however, to ensure that all loudspeakers have the same sensitivity, otherwise that with the highest sensitivity will be overloaded while the others are struggling to deliver their nominal power. This is also true, of course, of parallel-connected discrete output amplifiers. Fortunately, this sort of problem is easily prevented with the aid of a simple potential divider between mixer and active loudspeaker (or output amplifier). For example, two active loudspeakers have to be connected in parallel across the output of a mixer; the input sensitivity of speaker 1 is 2.5 V into 22 kΩ, and that of speaker 2 is 1.2 V into 50 kΩ. Clearly, speaker 2 is the more sensitive, since its input voltage is only 1.2 V for full drive of the amplifier. The input sensitivities must therefore be equalized. A potential divider can only reduce the sensitivity. So, the sensitivity of speaker 2 is to be lowered to that of speaker 1. This is done with the circuit in Figure 3.2.5. The value of resistor R_2 must be much lower than the input impedance of the loudspeaker. At the same time, this value must not be too low to avoid too heavy a load on the mixer. In prac-

136

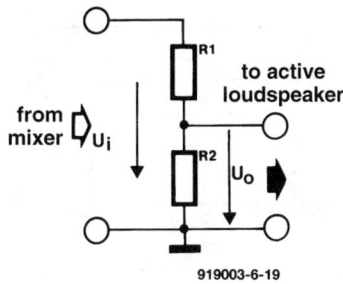

Figure 3.2.5. Potential divider to lower the input sensitivity of an amplifier.

tice, a good value of R_2 is $Z_i/10$. In this case, the load on the potential divider owing to Z_i can be ignored. The value of R_2 is then

$$R_2 = Z_i/10 = 50000/10 = 5000 \ \Omega.$$

Since this is not a standard value in the E-12 series, make R_2 = 4.7 kΩ (which is standard).

The value of R_1 is calculated by considering that at full drive, the voltage drop across R_2 is 1.2 V. The level of voltage provided by the mixer is as high as the input sensitivity of loudspeaker 1, that is, 2.5 V. The difference is dropped across R_1, so that

$$R_2/U_i = R_1/(U_{tot} - U_i)$$
$$\therefore$$
$$R_1/(2.5 - 1.2) = 4.7 \times 10^3/1.2$$
$$\therefore$$
$$R_1 = (4.7 \times 10^3/1.2)(2.5 - 1.2) = 5091.67 \ \Omega.$$

The nearest E-12 value is 4.7 kΩ.

To verify these values, consider that when there is a voltage at the input of the potential divider of 2.5 V, the potential across R_2 and therefore at the input of the active loudspeaker is

$$U_2/R_2 = U_{tot}/(R_1 + R_2)$$
$$\therefore$$
$$U_2 = U_{tot} R_2/(R_1 + R_2) = 2.5 \times 4700/(4700 + 4700) = 1.25 \ V.$$

The potential divider may be housed in the loudspeaker enclosure, but where this undesirable, it may be housed in a discrete metal case as shown in Figure 3.2.6. The screen of the con-

137

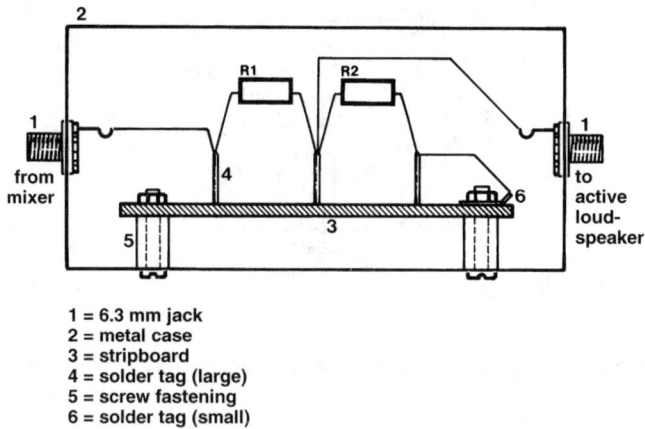

```
1 = 6.3 mm jack
2 = metal case
3 = stripboard
4 = solder tag (large)
5 = screw fastening
6 = solder tag (small)
```

919003-6-20

Figure 3.2.6. The potential divider shown in Figure 3.2.5 may be fitted in
a small, sturdy metal case, which is connected between the output
of the mixer and the input of the active loudspeaker.

necting cable must be strapped securely to the metal case to ensure good contact and to avoid any interference.

Some active loudspeakers have a rotary control or preset potentiometer at the back to enable the input sensitivity to be adjusted as required. Such a control is, however, of use only if it has a scale at which the set sensitivity can be read. Such a scale may be calibrated in V or in dB. If the scale is in volts, the sensitivity can be set without any problem; if it is in dB, some computation is necessary. In the earlier example, the input voltage to loudspeaker 2 must be 1.2 V for full drive. This means that the potential of 2.5 V must be multiplied with a factor

$$1.2/2.5 = 0.48.$$

According to Table 1.2.1, this corresponds to about 6 dB, which is the value to which the control must be set.

The cable connecting the mixer and active loudspeaker carries a small signal only and it must, therefore, be screened. As discussed before (2.1.3) a balanced connection is less sensitive to interference than an unbalanced one. Whether a balanced connection can be used depends on the active loudspeaker, however.

3.2.4 Single-channel installations

There are single-channel and two-channel voice installations on the market. Single-channel installations have only one output amplifier to which several loudspeakers may be connected. All signal sources connected to the mixer are reproduced by all loudspeakers at the same sound level. A simplified block diagram of such an installation is shown in Figure 3.2.7. The output voltage and internal resistance of the microphones are lower than those of most of the other signal sources, which is why the input sensitivity can be switched over at each input terminal. The switches enable the sensitivity of the amplifier to be matched to a microphone (mic) or to other signal sources (Line).

A well-designed voice amplifier has balanced microphone inputs. This is particularly beneficial when the amplifier is set up in a large room where the microphone cables are then fairly long. If at all possible, the amplifier is set up in the room in front of the stage, since the sound technicians are then able to keep an eye on, and adjust if necessary, the sound and/or volume of the voiceist(s) and the acoustic instrument(s).

The input signal is first applied to a preamplifier, in which the transfer factor can normally be adjusted with a gain control. This has the same effect as a continuously variable input sensitivity. This arrangement allows the amplification of virtually any kind of input signal. Also, the gain controls enable the levels of the various signals to be equalized, which makes operation

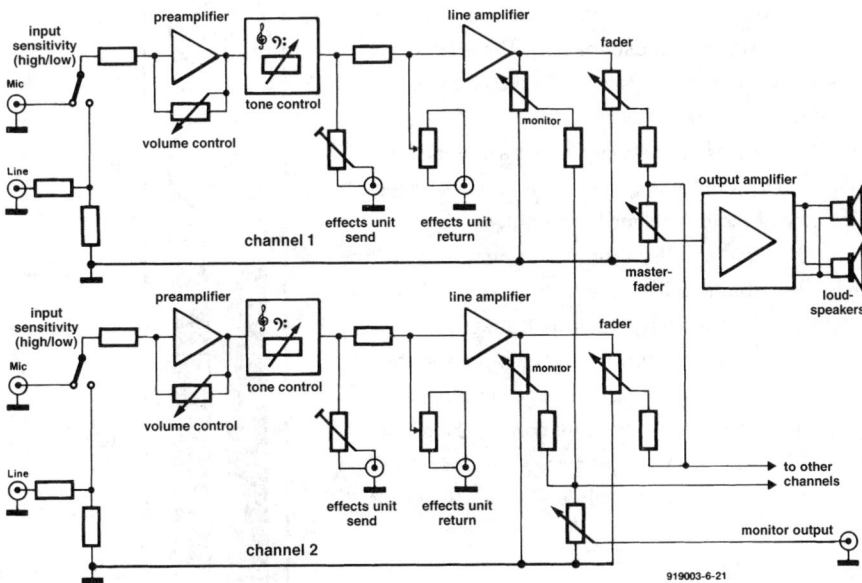

Figure 3.2.7. Simplified block diagram of a vocal amplifier with only one output amplifier.

139

of the mixer a lot simpler. The preamplifier is followed by a tone control which will be reverted to later.

Effects units if used are inserted between the tone control and the line amplifier. The control 'effect send' enables setting the level of the signal that is sent by the mixer to the effects unit; the 'effect return' control is used to set the level of the signal that is returnned by the effects unit to the mixer. This arrangement makes it possible for the degree of the effect to be controlled. On some equipment, the term 'effect' is replaced by 'echo' (echo send and echo return). However, any effects unit can be inserted into the signal path. Usually, the effect send level is set internally with a preset potentiometer.

The fader serves to control the volume of each channel. All channels are ultimately terminated into the output amplifier via the master fader with which the overall sound level is controlled.

Parallel to the main signal path leading to the output amplifier is a monitor signal path. With the aid of the associated controls, a separate output signal can be constructed which can be reproduced via a monitor amplifier and monitor loudspeaker(s). These speakers radiate into the direction of the voiceist so that he/she can hear what he/she is doing.

Usually, several loudspeakers may be connected to the output of the output amplifier. Care must be taken, however, that the total resulting loudspeaker impedance does not become smaller than the minimum output impedance of the amplifier (see also 1.1.9 and 3.1).

Figure 3.2.8 shows a typical front panel of a single channel of a small voice amplifier. The input sensitivity can be switched between mic(crophone) and line, while the gain control provides continuous control of the input sensitivity. The range of the gain control depends, of course, on the position of the mic/line switch. Controls

Figure 3.2.8. Typical front panel of one channel of a voice amplifier.

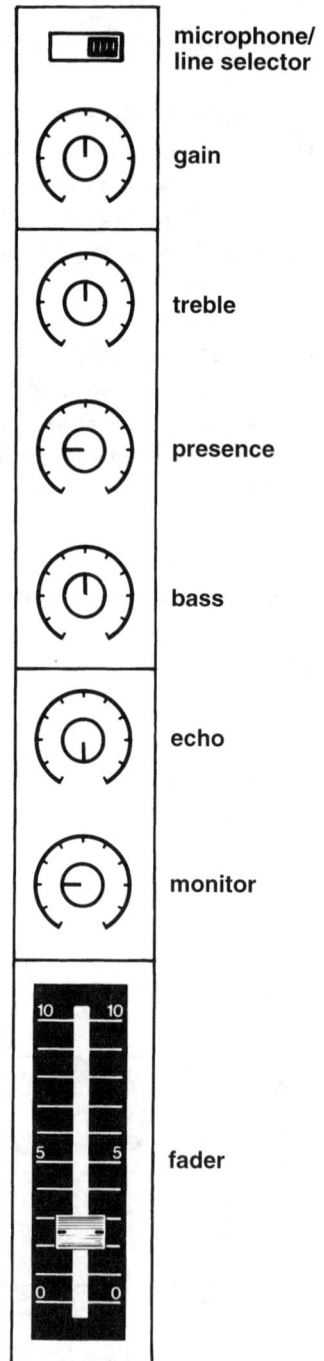

microphone/line selector

gain

treble

presence

bass

echo

monitor

fader

919003-6-22

140

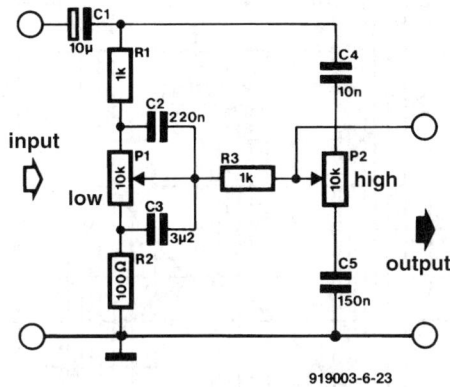

Figure 3.2.9. Circuit of a typical tone control that is found in many amplifiers. Basically, it consists of a low-pass section and a high-pass section.

treble, presence and bass belong to the tone control system. They are followed by the effects control and a monitor output control. The fader is, as customary, a slide potentiometer. This type of control is very convenient because it enables the operator to see at a glance what level is set. The longer the travel of such a control, the more accurate the setting. This is why many manufacturers include the fader length (that is, the length of travel) in the specification of the mixer.

There are various way of designing a tone control system. The simplest of these is to increase or reduce the level of a signal with reference to a fixed low and a fixed high frequency. This type of control is found in virtually all hi-fi equipment and also in simple voice amplifiers and mixers. The basic circuit of it is shown in Figure 3.2.9.

In the circuit, capacitor C_1 is an electrolytic component which must be connected with correct polarity at all times. Basically, the circuit consists of a variable low-pass filter and a variable high-pass filter (see 1.1.4). The effect on the frequency response of the amplifier is illustrated in Figure 3.2.10. The amplification of low frequencies (bass) and high frequencies (treble) can be adjusted independently of one another.

There are tone controls that allow the middle frequencies to be adjusted with a so-called presence control. The effect of this control on the frequency response is shown in Figure 3.2.11. The control allows the 'moving' of a soloist from the background to the foreground.

The simple tone control discussed does not suffice in more extensive mixers as used in public address installations and recording studios. In these a parametric equalizer is normally used, which will be reverted to in Chapter 5.

Figure 3.2.12 shows the front panel of a compact single-channel mixer. The simplicity of this unit makes it very interesting for a number of reasons. Particularly unusual is the setting of

141

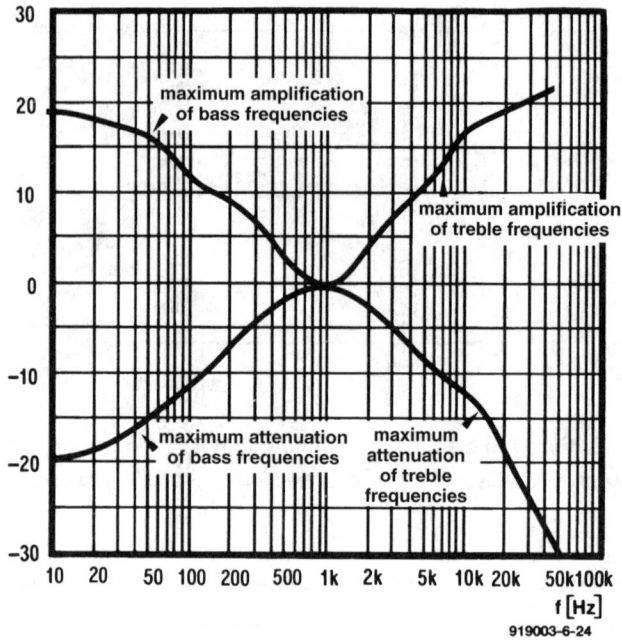

Figure 3.2.10. Illustrating the effect of a tone control system as in Figure 3.2.9 on the frequency response of an amplifier.

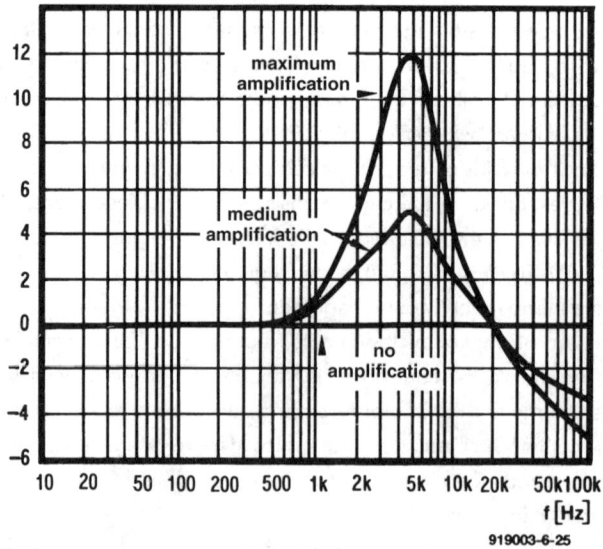

Figure 3.2.11. Typical frequency response curve of a presence control.

142

Figure 3.2.12. A small, compact power mixer with five input channels. There is tone control (treble and bass) and gain control for each channel. There is a facility for connecting an effects unit (rev) and a monitor (mon). The master section (OUT) enables the input sensitivity of the integral output amplifier and the signal level of effects and monitor to be adjusted. Three light-emitting diodes (LEDs) indicate whether the mixer is working correctly. Level Mon = output level of monitor; Power = on/off indication; Level PA = output level of the integral power amplifier.

the volume with the input sensitivity control (gain) instead of with (more expensive) faders. This is not only an inexpensive solution, but it also reduces the number of operating controls. It has a drawback, however: a fader reduces the noise added the signal by the amplifier to the same extent as the signal itself. This is a kind of noise suppression: at high volume, a large part of the signal and the internal noise are applied to the output amplifier, while at low volume, the signal as well as the internal noise are attenuated. With a gain control, the internal noise is not attenuated, so that at low volume the signal-to-noise ratio is degraded.

The situation just described is represented by the graphs in Figure 3.2.13. The upper trace shows what happens when the volume is decreased with a fader at the output of the preamplifier. It is seen that both the amplitude of the signal and the noise level are reduced. When the volume is turned down with a gain control, which precedes the preamplifier, the noise level is not reduced and becomes actually quite audible at certain settings of the gain control until the sig-

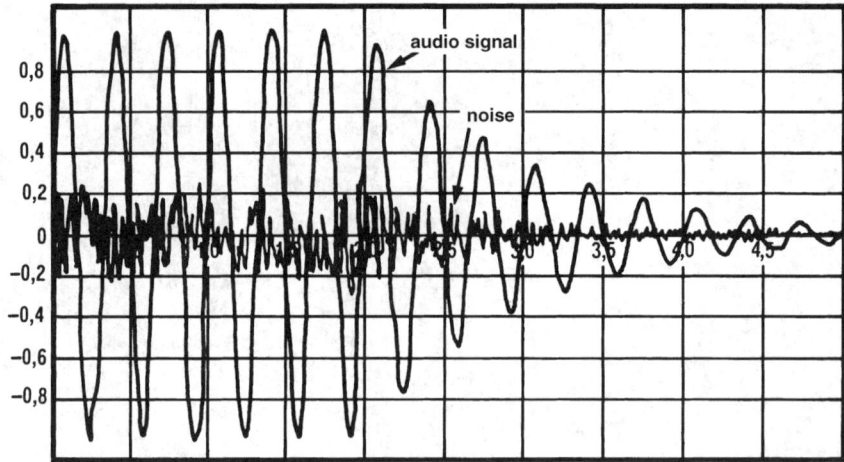

919003-7-2

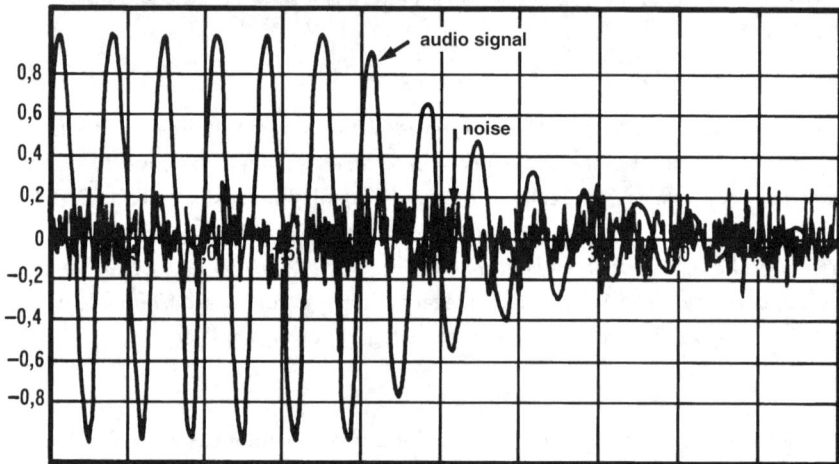

919003-7-3

Figure 3.2.13. These graphs illustrate why a gain (input sensitivity) control, preceding the preamplifier, is not suitable for adjusting the volume. Even when it is fully turned down, the internally generated noise is reproduced unattenuated (bottom). A fader, following the preamplifier, attenuates the signal and the internal noise to the same extent (top).

nal is submerged by the noise and only noise is heard.

As an aside: equipment operating with operational amplifiers (op amps) instead of discrete transistors often generate much internal noise, although this is not so if modern low-noise op amps

144

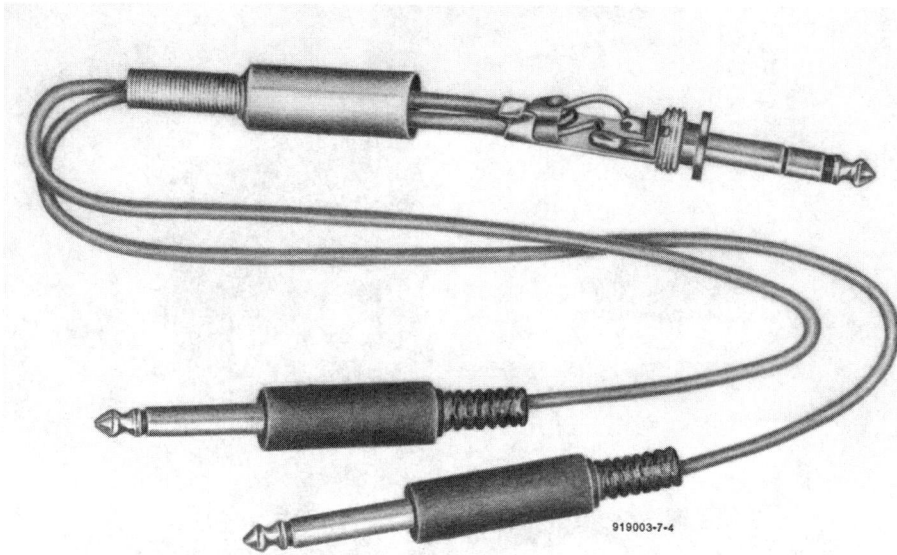

919003-7-4

Figure 3.2.14. Cables for connecting an effects unit to the mixer in Figure 3.2.12. The stereo 6.3 mm audio plug (shown opened) is used with many voice amplifiers and mixers. Since most effects units have isolated input and output sockets, two separate (mono) plugs are needed for connection to the effects unit. The screen of the two cables is soldered to the housing of the plugs.

are used.

Otherwise, the unit has all the facilities needed: two-fold tone control (bass and treble), monitor output, socket for connecting an effects unit, and an amplifier with a nominal output of 100 W, to which two 8 Ω loudspeakers can be linked. The output power is shared equally by the two loudspeakers. There is also a line output where a signal taken from just before the output amplifier is available. An effects unit must be connected with a cable as shown in Figure 3.2.14.

All inputs are intended for low-impedance microphones (Z_i = 600 Ω). High-impedance instruments, such as an electric guitar, must be connected via an impedance transformer as described in 2.1.3. There are only two indicator LEDs: one for the main signal path and the other for the monitor output. When a signal is available at that output, the LED lights.

There is one special facility: a multi-way socket to which all inputs are connected has been added after manufacture. This socket can accept a multicore cable containing all the necessary leads between the mixer and the stage, which makes a much neater appearance (and is safer, too). All microphones and instruments that are to be reproduced via the mixer are con-

919003-7-6

919003-7-7

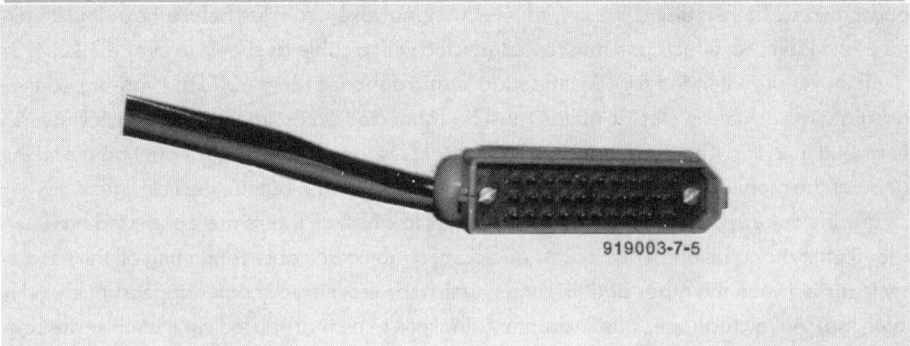

919003-7-5

146

nected to a stagebox that is linked to the mixer via a multicore cable (see Figure 3.2.15). Bear in mind that unbalanced lines must not exceed a length of about 10 metres (33 ft), including the lines on the stage to prevent the risk of their picking up noise and hum. The stagebox in the photograph has, apart from input sockets for the microphones, a socket for the monitor output with associated control (potential divider), which enables the monitor to be controlled from the stage.

3.2.5 Two-channel installations

For live performances of rock music in a small hall, it is normally not necessary to use a full-blown public-address system. The basic instruments for rock music (guitars and keyboards) are usually noisy enough without amplification, but the percussion may need some. In such a case, it is perfectly feasible to use the voice amplifier(s). It is then, however, necessary to use a two-channel system with sufficient spare output power. Such a system contains two isolated output amplifiers which share the various input channels.

Each instrument (input channel) can be placed at random in the sound spectrum between the two loudspeakers with a so-called pan-pot. This makes the sound spectrum clearer and more transparent than is possible with a single-channel system – less is lost from the original, natural sound. A pan-pot has one input and two outputs which are linked to the loudspeakers. Such a control can be designed in various ways, three of which are shown in Figure 3.2.16.

When the pan-pot is at the centre of its travel, both amplifiers are supplied with equal signal levels. When it is turned fully clockwise or fully anti-clockwise, however, the signal is applied to one of the amplifiers only. In between these extreme settings, the amplifiers are supplied with different signal levels, resulting in a different position in the sound picture. Typical characteristics illustrating the operation of a pan-pot circuit are shown in Figure 3.2.17. Note that both channels are attenuated by 3 dB when the control is at the centre of its travel. Since the sound from the two loudspeakers is combined when we listen to it, we hear the same sound level whatever the position of the pan-pot. In other words, turning the pan-pot changes the direction from which the sound seems to come, but not its intensity.

Figure 3.2.18 shows the simplified block diagram of a two-channel voice amplifier. For

Opposite page: Figure 3.2.15. A receptacle for a multi-core cable enables all connections between mixer and stage to be contained in one cable. Such a receptacle can be added to mixers that have not already got one. Note that unbalanced lines must not be longer than 10 metres (33 ft) including the microphone cable on the stage to prevent them picking up a large amount of noise and hum. The connection in the illustration contains, apart from the lines to all inputs, also a line to the monitor output so that musicians on the stage can set the desired monitor volume themselves.

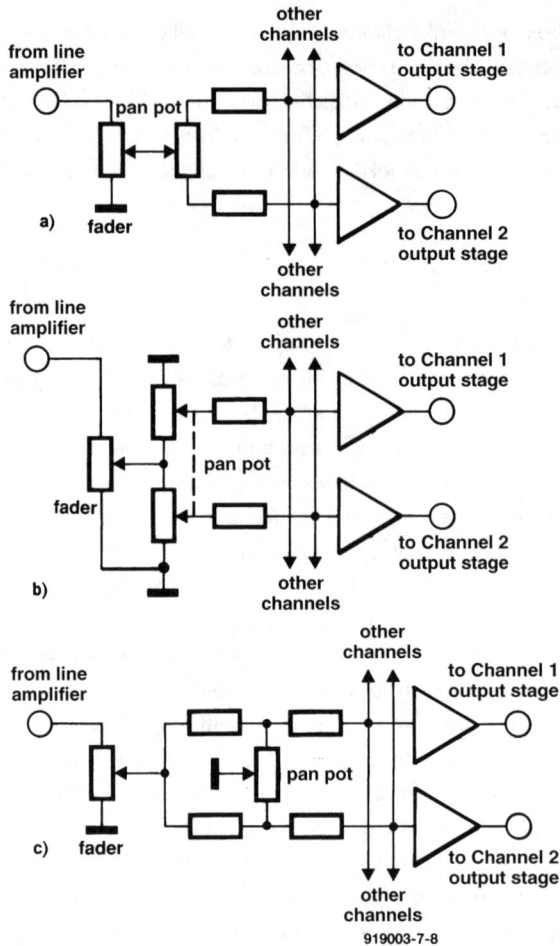

Figure 3.2.16. A number of variations of the pan-pot. They all work in the same way, however: the signal can be shifted seamlessly between the two channels (left-hand and right-hand).

clarity's sake, the monitor outputs and the master have been omitted. The main difference between this and Figure 3.2.7 is the two isolated output amplifiers. Each channel has its own pan-pot which enables the seamless shifting of the input signal from extreme left to extreme right.

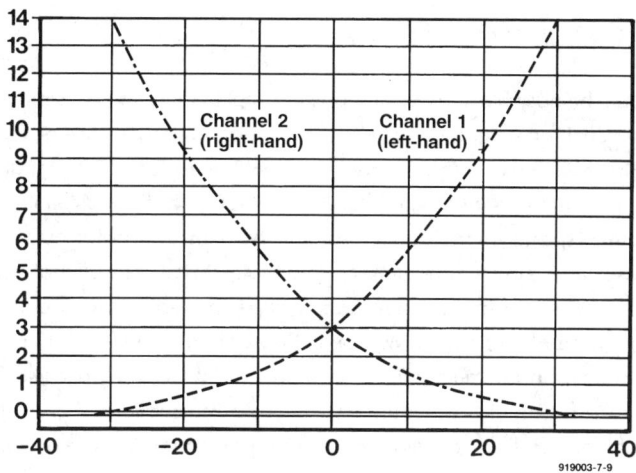

Figure 3.2.17. Curves showing the attenuation of the signal in both channels as a function of the position (angular rotation) of the pan-pot.

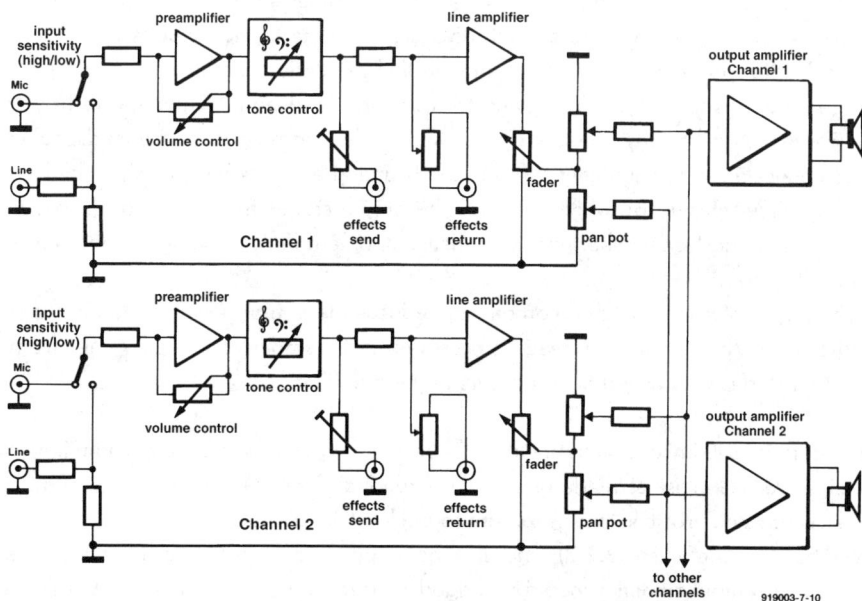

Figure 3.2.18. Simplified block diagram of a two-channel voice amplifier.

149

3.2.6 Output amplifiers

In principle, it would be possible to use a hi-fi power amplifier in a voice installation with passive loudspeakers. Such an arrangement is, however, not advisable since as far as practical application is concerned, the two designs are incompatible. As far as transfer and sound are concerned, there may not be that much difference, although hi-fi amplifiers generally have slightly better electrical specifications than voice amplifiers. However, hi-fi amplifiers are intended for domestic use, where the temperature and humidity do not vary all that much throughout the year. Also, they are normally in a more or less permanent position so that the various connectors are not used all that often. Moreover, they are not intended to deliver maximum power for long periods. Finally, mechanically they are not robust enough to withstand continual transportation. Professional power amplifiers for voice and public-address installations are virtually indestructible. Parts like the mains power supply and heat sinks are designed for long-duration heavy loading. Moreover, they have integral safety circuits such as:

- DC protection, which ensures that it is impossible for direct voltage components to appear at the output of the amplifier. Direct voltages destroy loudspeakers within a very short time. If a direct voltage should nevertheless be present in the output signal, the protection circuit immediately disconnects the amplifier from the loudspeaker(s).
- Power-on delay, which ensures that the supply voltage to the output stages rises only slowly after power-on. When the supply to the installation is switched on, a large voltage peak ensues, which, if allowed to reach the loudspeakers, would destroy small drive units, particularly tweeters. The power-on delay arranges for the loudspeakers to be connected to the amplifier a few seconds after the supply has been switched on.
- Short-circuit protection which prevents the output stages from being destroyed when there is an inadvertent short-circuit at their output terminals during the setting up of the installation.
- Thermal protection, which automatically reduces the output power when the temperature inside the amplifier exceeds a preset level or removes the supply lines from the output stages when overheating occurs in the amplifier.

Tone controls and similar circuits are totally superfluous in professional output amplifiers: they merely increase the risk of interference and operational errors. Nevertheless, most amplifiers have a volume control to set the maximum sound level. In its stead, an input sensitivity control would be more useful (see 3.2.3), since this makes possible correct matching to the mixer. In principle, an output amplifier should be judged on its robustness and reliability. Even the simplest of amplifiers meets all electro-acoustic requirement.

When considering the power output, the discussions in 1.2 and 1.3 should be borne in mind. If only voice(s) and perhaps an acoustic instrument like a saxophone or recorder need to be amplified, the sound level of the percussion section should be taken as reference. To reach or supersede that level, both the amplifier and the associated loudspeaker(s) must be capable of so doing. A good minimum power output is 100 W. If the amplifier is also used, from time to time, to amplify percussion instruments, that power is not sufficient: it should be at least twice as high. And if now and then a keyboard or double bass has to be amplified, the power output must be larger still. However, exceptionally large output powers are not recommended, since amplifiers providing such powers are not only quite expensive, but invariably blow the fuses when they are used with normal domestic mains supplies.

3.2.7 Spurious sounds

A sound installation does not always provide joy: sometimes, it emits an annoying hum or buzz or even the programme of a nearby transmitter. Basically, there are two kinds of interference:

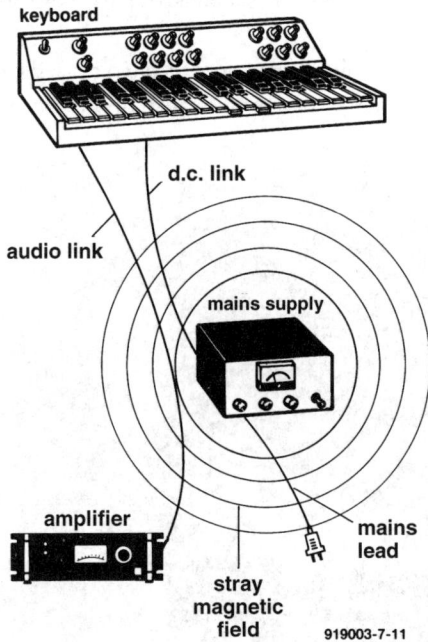

Figure 3.2.19. Power supplies of keyboard instruments and effects units produce alternating magnetic fields (stray fields) that are strong enough to induce a hum voltage in nearby audio connectors. Such equipment should therefore be kept away from signal-carrying leads.

- those caused by electromagnetic radiation (not connected with the mains supply);
- those reaching the installation via the mains supply.

Electromagnetic interference is caused by nearby transmitters, defect electrical equipment, mains transformers and inadequately decoupled computers. All these sources radiate electromagnetic waves that induce interference voltages in long microphone and loudspeaker cables, particularly when the sound installation consists of discrete units. Power mixers are far less vulnerable to such interference. Cables carrying small signals (such as microphone cables) are particularly prone to pick up outside interference and should, therefore, be screened types.

Balanced connections are far less sensitive to electromagnetic radiation than unbalanced ones. When interference occurs, the first thing to be checked is the proper screening of cables where appropriate. Even screened cables should not be laid parallel to power supply and loudspeaker cables. Mains adaptors for the supply of keyboard instruments or effects units must be kept away as far as possible from signal-carrying leads. Such adaptors use a mains transformer that produces a strong alternating magnetic field (stray field), which can induce noise voltage in a signal-carrying lead, which becomes audible as a stubborn 50 Hz mains hum (see Figure 3.2.19).

Noise reaching the equipment via the mains supply is often caused by defective or insufficiently decoupled circuits, earth loops or mains hum. Such noise does not consist of electro-

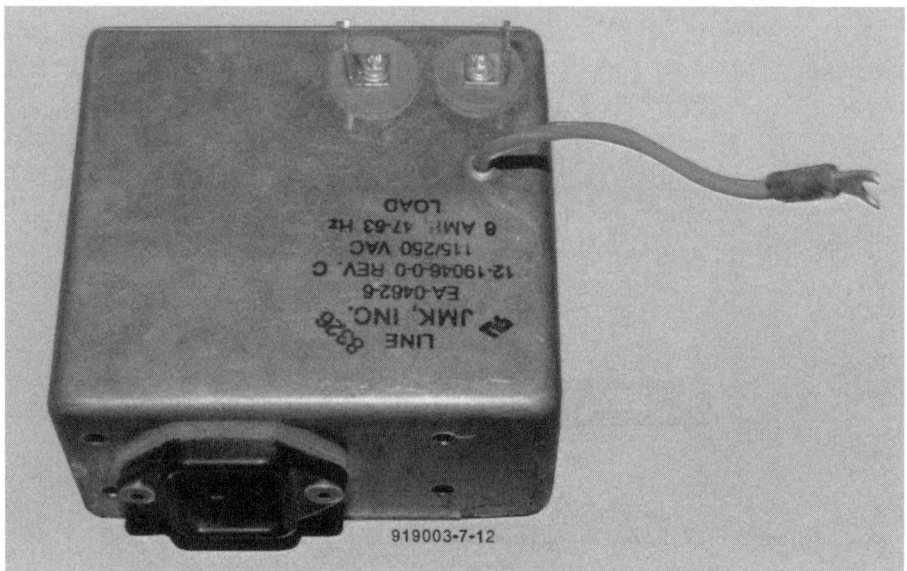

Figure 3.2.20. Mains filter for suppressing noise voltage. This kind of filter is often built in as standard in much equipment, but it may be added at a later stage if necessary.

magnetic waves, but is dispersed via connecting cables, particularly those connected to the mains supply. Large amplifier installations therefore normally incorporate a special mains filter as shown in Figure 3.2.20 to suppress such noise at the mains entry. In stubborn cases, a wire reel may be used (see Figure 3.2.21).

The second kind of noise voltage in the leads is caused by so-called earth loops, which are sometimes very difficult to trace and eradicate. They often result from the fact that the mains earth is connected to the metal enclosure of the various units in the sound installation. Often, the circuit earth, and therefore the screening of various cables, such as microphone cable, is also connected to the metal enclosure. So, there is an electrical link between the amplifier earth and the mains earth. That link is essential, but it must be made at only one point in the sound installation. In the case of discrete units that are individually plugged into the mains supply as in Figure 3.2.23, earth loops are unavoidable if unbalanced interconnections are used. The

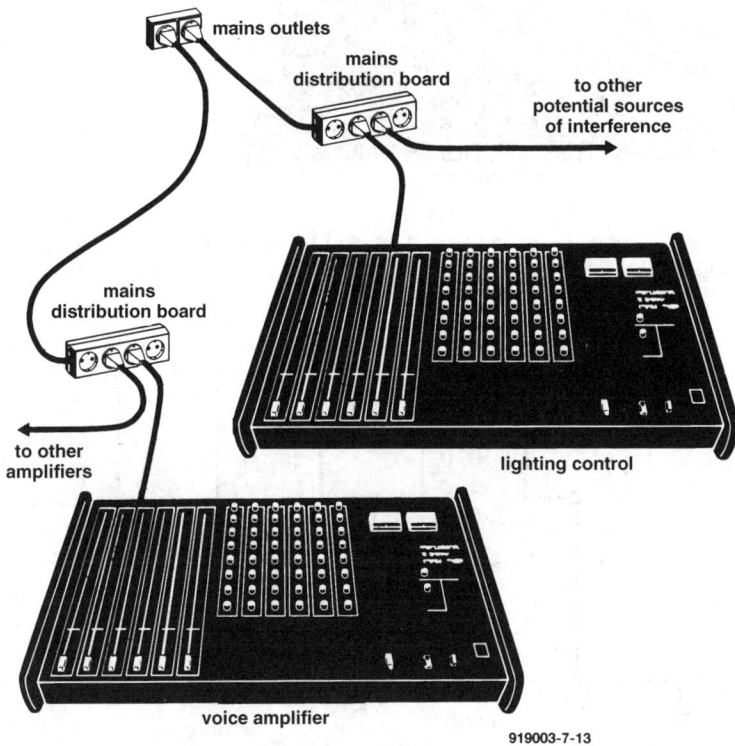

mains outlets

mains distribution board

to other potential sources of interference

mains distribution board

to other amplifiers

lighting control

voice amplifier

919003-7-13

Figure 3.2.21. The voice amplifier and possible sources of interference, such as smoke alarms, should, if at all possible, be plugged into different mains outlets
If that is not feasible, the mains leads should be kept well separated.

153

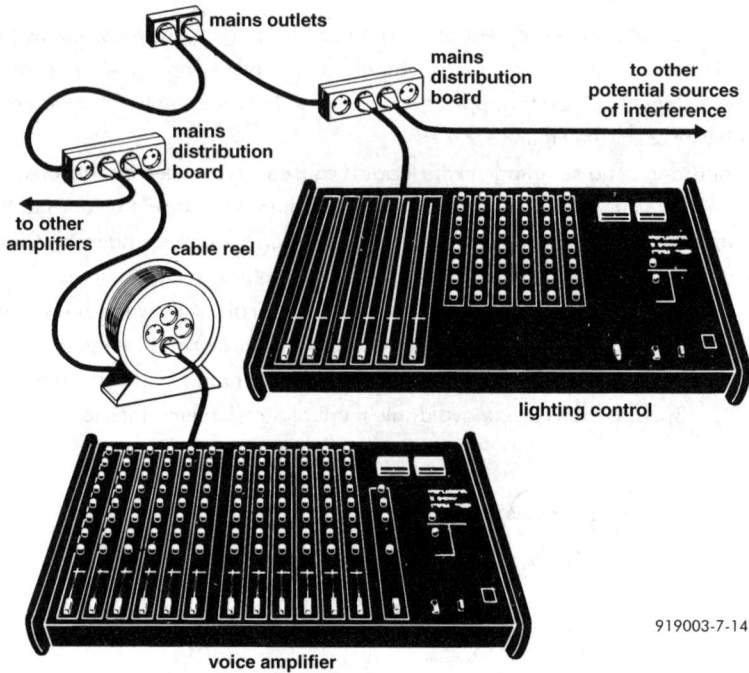

Figure 3.2.22. In many cases, noise pick-up can be prevented by inserting a reel of mains cable in the mains supply connection (which therefore becomes much longer).

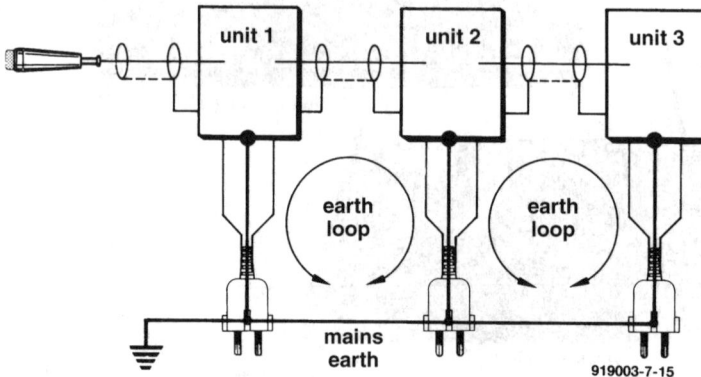

Figure 3.2.23. When a number of earthed units with unbalanced inputs are interconnected, earth loops may ensue.

only practical guaranteed way of getting rid of earth loops is to use balanced interconnections between the various units. It is, of course, a prerequisite that the various inputs and outputs of the units are suitable for this. In balanced connections, the signal-carrying leads are always isolated from the cable screen and therefore from the signal earth. Earth loops can therefore not ensue. It is, of course, possible that the cable screen picks up hum and other noise, but this cannot get into the signal path. Protection against earth loops and the picking up of hum can be improved by earthing only one end of the cable screen, since this then cannot pick up any hum or noise. See Figure 3.2.24.

The third kind of mains hum is caused by transformers and other inductors whose stray fields

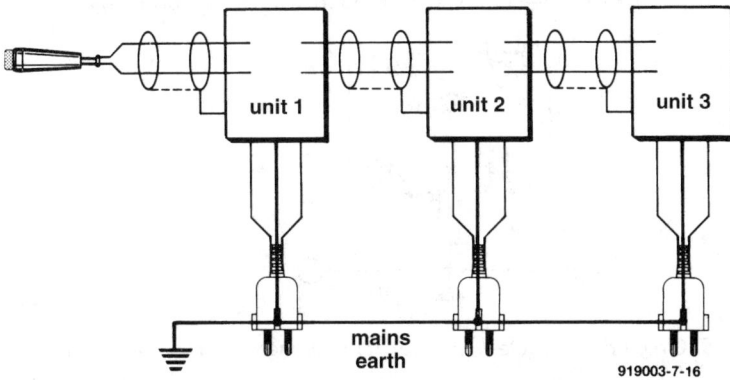

Figure 3.2.24. Balanced interconnections between a number of units prevents earth loops, particularly when the cable screens are earthed at one end only.

induce a noise voltage in the signal-carrying leads, or by equipment whose mains supply is defective, not well designed or inadequately screened. Mains hum can be recognized readily since its frequency is always exactly 50 Hz or 100 Hz (in some countries, such as the USA and Canada, 60 Hz or 120 Hz). Equipment in good working order should not suffer from mains hum. This is not always true of home-produced equipment, effects units that operate with an incorrect power supply, or older equipment. In short: mains hum is easily prevented by the use of correct equipment and adjuncts.

If a source of noise cannot be traced or eradicated, the only solution lies in suppressing the interfering frequency or frequencies. Thus may be done with the aid of an equalizer, which is capable of suppressing narrow frequency bands, or with a so-called double-T filter. The simplest design of such a filter consists of some resistors and capacitors to suppress a single frequency. It should be borne in mind, however, that both these solutions are emergency ones:

they do not eradicate the noise source. Also, they do affect the frequency response of the relevant installation so that the operation of this is no longer one hundred per cent.

In practice, it often occurs that older equipment, such as an ancient echo unit, causes interference. Frequently, a double-T filter at the output of such a unit can improve matters appreciably. A design for a filter to suppress mains (50 Hz) hum is shown in Figure 3.2.25. The transfer function of the filter is

$$A = (1 - x^2)/\sqrt{[(1 - x^2)^2 + 16x^2]},$$

in which $x = 6.24fRC$, where f is the frequency of the interference. With values as specified in Figure 3.2.25, the response of the filter is as shown in Figure 3.2.26. The 50 Hz mains hum is

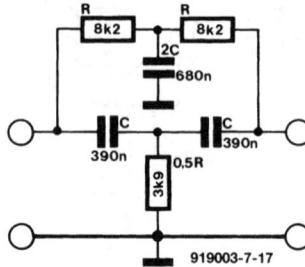

Figure 3.2.25. Circuit of a double-T filter for suppressing a 50 Hz mains hum.

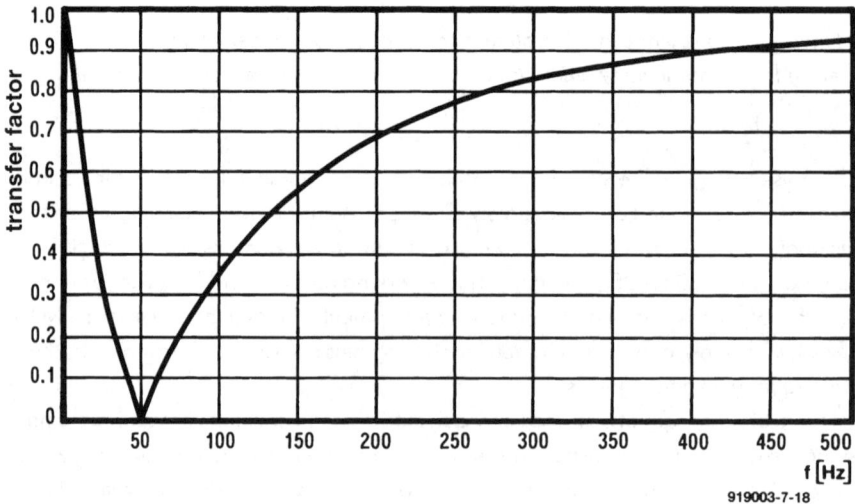

Figure 3.2.26. Frequency response characteristic of the filter in Figure 3.2.25.

156

suppressed almost completely. Unfortunately, adjacent frequencies are also attenuated, and this will have some effect on the reproduced sound.

The filter may be housed within the relevant unit or in a separate small metal case. In the latter case, the filter can be inserted into the signal path of a variety of units by means of two audio plugs and sockets. The connecting cables should be screened, with the screen firmly strapped to the metal case.

3.2.8 Setup of a voice installation

Even a good-quality voice installation needs to be positioned correctly in a room to give good results. Criteria for a correct position are:

* negligible tendency to howling;
* optimum sound distribution in the room or hall;
* good control over the sound.

To meet the first requirement, the loudspeaker(s) of the voice installation and the monitor loudspeaker must be considered quite separately. The former should be placed so that the sound they radiate cannot be picked up by the microphone(s). The risk of this happening all the same is great since the microphones are placed at the front of the stage. However, the loudspeakers can normally be placed so that they radiate directly into the room or hall. Positioning the monitor speakers is rather more difficult: these should be placed so that they point to the rear of the microphone, which therefore should preferably have a cardioid polar diagram—see Figure 3.2.27. When the vocalist takes the microphone from its stand, he/she must be careful not to

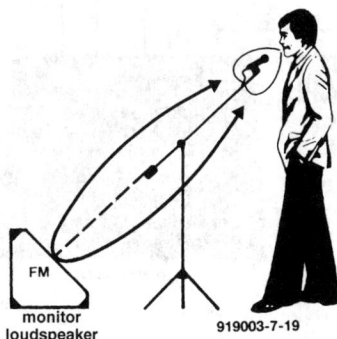

Figure 3.2.27. The monitor loudspeaker must be placed in such a way that as little sound as possible from it is picked up by the microphone. This is why microphones with a cardioid polar diagram are particularly suitable in this situation.

157

point it at the monitor speaker(s).

To cover the room or hall well, the polar diagram of the loudspeakers should be borne in mind. Normally only the vocal parts and acoustic instruments are reproduced via the voice installation—sometimes part of the percussion section. The sound of the voice loudspeakers must merge in well with the sound of other instruments. Also, they should not radiate the sound straight ahead, but should be pointed slightly towards one another as shown in Figure 3.2.28. This arrangement prevents an apparent 'hole' in the sound at the centre of the room or hall which causes the vocal parts to be heard there badly or not at all.

During live performances, the mixer or voice amplifier should be placed in the hall in front of the stage, if possible at the centre. It is at that location that the sound technician is in the best position to hear the results of his/her actions. Unfortunately, not all voice installations have balanced inputs, so that the length of the microphone cables must be kept short. To keep all the cables tidy, a stage (intercabling) box (home made) is often used to which all microphones and

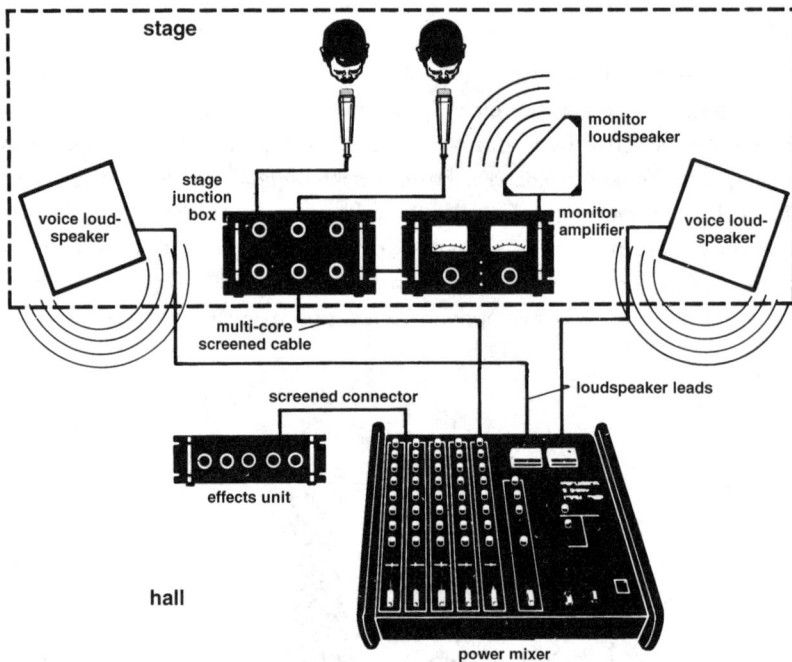

919003-7-20

Figure 3.2.28. Correct positioning of a voice installation during a live performance. The mixer must be placed in the hall in front of the stage to enable the sound technician to monitor the sound in the hall and to set the sound level of the vocal parts and any instruments amplified via the installation as required.

158

instruments are connected. Only a multicore cable is then needed to link the box to mixer and to the monitor speaker.

3.3 Musicians' amplifiers

Musicians' amplifiers (also called musical instrument amplifiers) are used primarily for amplifying electro-musical instruments and so fulfil the function of the sound board in acoustic instruments. These amplifiers do not merely amplify the sound, they also shape it and so influence the timbre of the instrument(s). So, the amplifier has a sound of its own, and this applies particularly to a guitar amplifier. As far as a musician is concerned, his/her guitar (or other instrument) is only part of the instrument: the whole instrument is the combination of the guitar, amplifier and loudspeaker.

3.3.1 General

A common property of musicians' amplifiers is their robustness. Good-quality amplifiers are constructed on a strong, steel chassis with additional mechanical reinforcements. the chassis of some models is protected against corrosion by cadmium plating. The enclosure is usually made of thick, solid wood. Leatherette cladding and steel corner braces prevent scratching.

Electronically, there are two kinds of musicians' amplifier: valve and transistor. Those using operational amplifiers (op amps) or other integrated circuits (ICs) perform generally as transistor amplifiers. Since an IC contains a large number of transistors, they normally produce more internal noise than transistor models. Valve amplifiers are used nowadays if sound shaping by the amplifier is required: the frequency response and distortion are of no significance. Transistor amplifiers are used when a straight frequency response is needed and no or very special distortion is wanted.

A comparison of the two types immediately shows the difference in size. Compared with that of a transistor amplifier, the design of a valve amplifier is very simple. There are two reasons for this: (1) in valve amplifiers, which do not need hi-fi quality, the amplification of each valve can be very large so that only few stages are needed, and (2) a valve amplifier does not need much protection since valves in general can cope very well with overloads.

The operational reliability of the two kinds of amplifier are also very dissimilar. The output stages of a musicians' amplifier must be be proof against short-circuits and overloading (for instance, resulting from too low a loudspeaker impedance in combination with high drive powers). Older transistor amplifiers frequently gave serious problems in this respect: they were apt to give up the ghost in a fraction of a second. This is, of course, particularly embarrassing during a live performance. Moreover, such an amplifier can normally not be repaired easily or in situ, coupled with the fact that older type transistors are often no longer available so that equivalent types have to be found, which often proves impossible.

Valve amplifiers are much better capable of coping with short-circuits and overloads. Also, if a valve should give up the ghost, it can be replaced in a jiffy, even by a layman (see Chapter 9). Normally, there is no problem in connecting a number of loudspeakers to one (valve) amplifier, since the exact value of the loudspeaker impedance does not matter all that much. On the other hand, open-circuit outputs (no loudspeaker or other load connected to the amplifier output) are not taken so kindly and may damage the output valves. At maximum drive, it may even happen that voltage flash-over occurs in the output transformer, which is then destroyed. Note that this cannot happen when the output stages are fed back via the secondary winding of the output transformer. Replacing an output transformer is a tedious and costly job, ignoring the fact that the correct type is often no longer available.

The most reliable are modern solid-state output amplifiers with extensive protection circuits (see 3.2.6) and integral current limiting circuits. These amplifiers are proof against overload as well as against open circuits.

The inputs of these power amplifiers are just as interesting as the outputs. Many amplifiers are fitted with two switched input sockets per channel, which enable one of two different input sensitivities to be selected. The ratio of these two is usually 1:2. Since different instruments invariably have different signal output levels, this facility is very useful. When the two inputs are used simultaneously, they have the same input sensitivity. This arrangement is not advisable since the inputs are internally interlinked. The volume and tone controls of the relevant channel then influence both instruments, which may be very inconvenient.

3.3.2 Guitar amplifiers

A guitar amplifier as shown in Figure 3.3.1 is intended specifically for use with an electric guitar: today, special amplifiers for acoustic guitars are very rare. In any case, an acoustic guitar, particularly the classical guitar, which has to be reproduced as faithfully as possible, may be amplified via the voice installation. Of all musicians' amplifiers, the guitar amplifier is the one that contributes most to the final sound of the instrument. Faithful reproduction does not come into question, because the sound heard by the listening audience is mainly the product of the amplifier.

In the final analysis, only the sound produced by the combination of guitar, amplifier and loudspeaker is of importance to the listener(s). The loudspeaker associated with the amplifier is the sound source whose output may be processed further for recording or reproduction via amplifiers that provide a faithful reproduction. Because of this, the frequency response of a guitar amplifier may be curved as in Figure 3.3.2 and its distortion factor may be accordingly high. The same applies, albeit to a slightly lesser extent, to other musicians' amplifiers. The design of a guitar amplifier therefore requires not just knowledge of electronics and transfer technology. A designer of a guitar amplifier is more like an instrument designer, which is the

Figure 3.3.1. Type Jive guitar amplifier from Engl.

reason why the circuits of guitar amplifiers differ basically from those of hi-fi amplifiers.

Figure 3.3.2 shows the transfer characteristics of a guitar amplifier with the two extreme settings of the tone control. The curves show only a part of the signal changes: only the linear distortion is incorporated. A guitar amplifier must also produce non-linear distortion, however, and this is provided by the active components in the circuit: valves, transistors, diodes and/or ICs. Typical of non-linear distortion is that harmonics are added to the signal that were not present in the original signal. The frequency and amplitude of these harmonics are of great importance to the final sound.

Since the properties of valves and transistors are quite different (they are normally given in the form of diagrams), these components produce different harmonics (sometimes called transients). Consequently, a valve guitar amplifier sounds basically different from a solid-state guitar amplifier, but this is true only, of course, if the amplifier is actually designed for non-linear performance. For convenience's sake let it be assumed that a valve amplifier has a quadratic characteristic and a solid-state amplifier an exponential one. If a pure sinusoidal signal is represented on a quadratic scale, harmonics as shown in Figure 3.3.3 ensue, whereas if it is represented on an exponential scale, the harmonics shown in Figure 3.3.4 result.

161

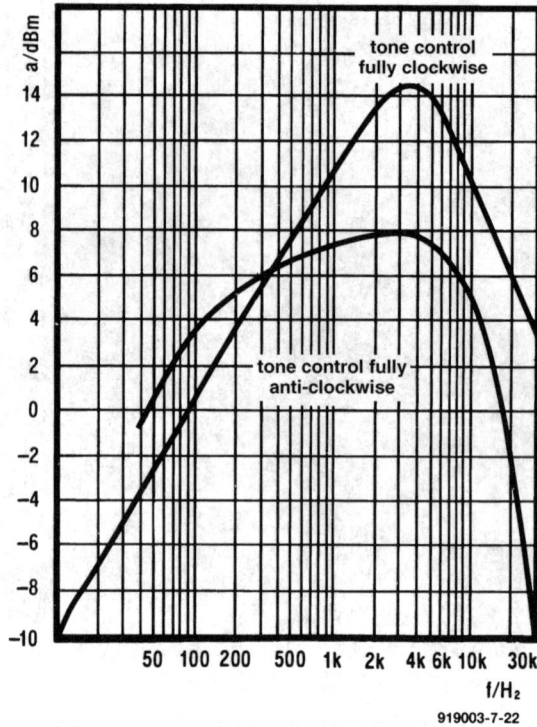

Figure 3.3.2. Frequency response characteristic of a small valve guitar amplifier.

In Figure 3.3.3, a direct voltage component, which is of no further interest here, and a signal at twice the frequency of the input signal ensue. In Figure 3.3.4, again a direct voltage component that is of no further interest, and two signals at twice and treble the frequency of the original input signal result. Obviously, this has an appreciable effect on the timbre.

Another important difference between solid-state and valve amplifiers is that the latter are intentionally overdriven to some extent. The greater the overloading, the more the final signal tends to become rectangular. In solid-state amplifiers the resulting waveform has sharp corners, whereas in valve amplifiers these corners are rounded. An analysis of these two types of waveform—see Figures 3.3.5. and 3.3.6—shows that the peak values of the sharp-cornered one are greater than those of the rounded-corner waveform. Clearly, this leads to a quite different timbre.

Although the foregoing theoretical approach is in practice tenable only to some degree, it makes clear that different types of amplifier design, and particularly what kind of active components are used, produce a different output signal. However, in a valve amplifier not only the

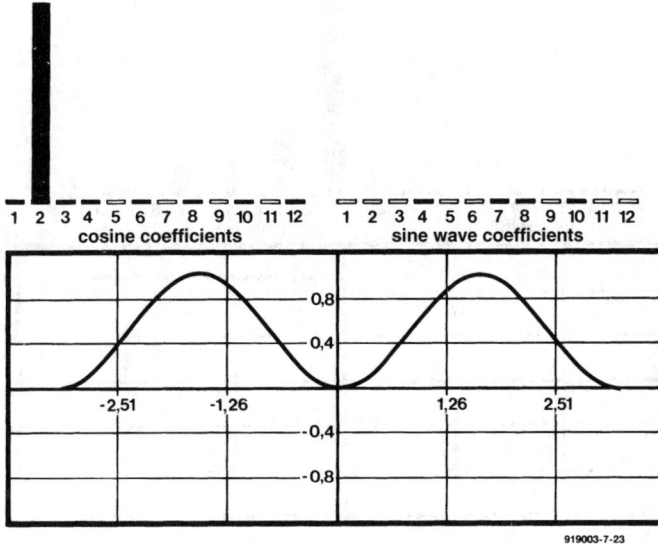

Figure 3.3.3. When a sinusoidal signal is represented as a quadratic function (the simplified transfer function of a thermionic valve) harmonics ensue.

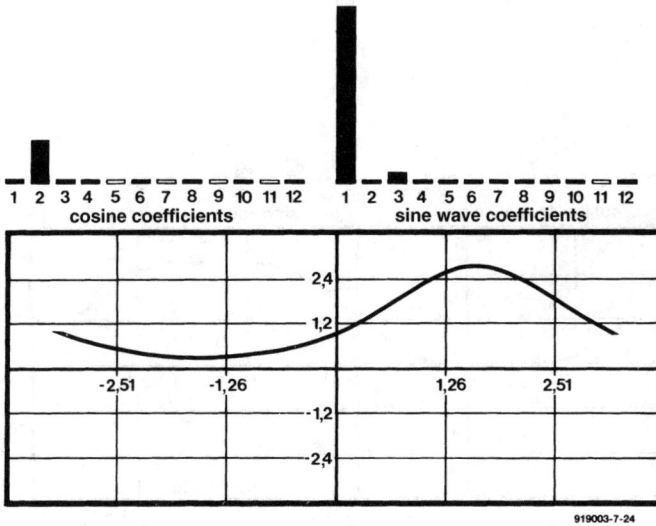

Figure 3.3.4. When a sinusoidal signal is represented as an exponential function (the simplified transfer function of a transistor) different harmonics than in Figure 3.3.3 result.

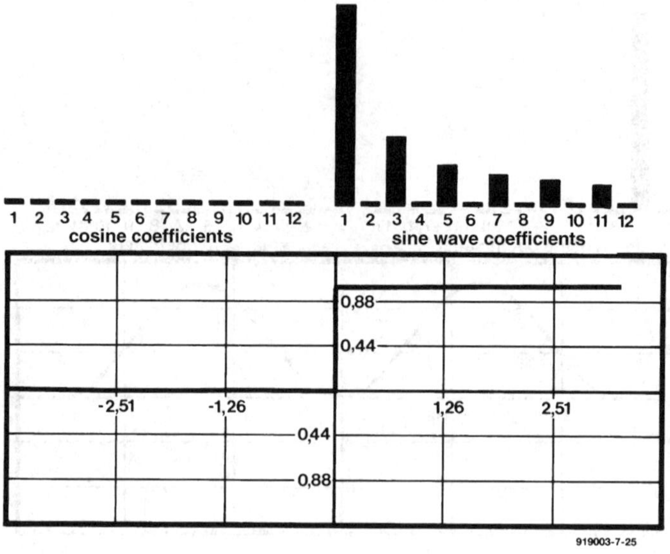

Figure 3.3.5. Illustration of the analysis of a square-wave signal into sine and cosine coefficients. Note the large number of harmonics.

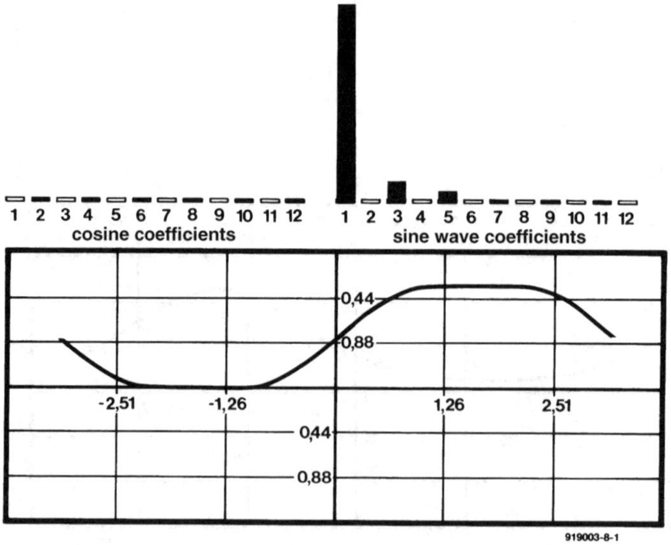

Figure 3.3.6. When a rounded square-wave signal is analysed, the peak values of the harmonics are much smaller than in the analysis in Figure 3.3.5.

valves, but also the output transformer, have a pronounced effect on the final output signal. At high levels of overdrive, the iron core of the transformer becomes saturated, which causes non-linear distortion of the output signal. The transformer also produces linear distortion.

Apart from harmonics, the dynamic range of the amplifier also affects the sound output. As described earlier, the usable dynamic range is limited at its lower end by the internal noise of the amplifier. An exact upper limit of the dynamic range of a guitar amplifier cannot be given, since this depends on how much distortion is acceptable to the guitarist. For the same reason, the power output of a guitar amplifier can only be stated approximately since many manu-facturers do not abide by the measuring standards discussed in 3.1. Figure 3.3.7 shows the basic distortion factor, k, as a function of power output, p. In a solid-state amplifier, the distor-tion rises rapidly from a certain output level (curve a), whereas in a valve amplifier it increases gradually (curve b). This means that a customarily overdriven valve amplifier has much more spare output power capacity than a similarly overdriven transistor amplifier, which explains why the dynamic ranges of these amplifiers are different.

There are other causes for the different dynamic responses: (1) the dissimilar slopes of the transfer characteristics of valves and transistors, and (2) the very different operating voltages.

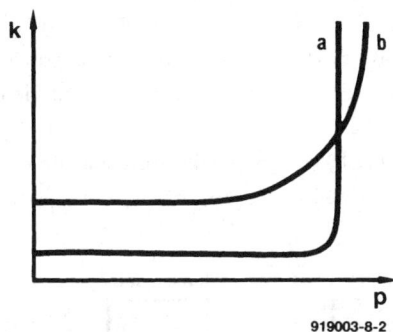

919003-8-2

Figure 3.3.7. The performance of an overloaded valve amplifier is clearly different from that of a similarly overloaded transistor amplifier. In the valve amplifier, the distortion factor increases only gradually, whereas in a solid-state amplifier it rises abruptly above a certain power output.

At a certain output, the relatively low supply voltage of transistor circuits (9–24 V) becomes an impassable barrier to further increases in the peak value of the output signal. The high supply voltages of valves (300–500 V) enable very high peak values to be processed. It must be restat-ed again, however, that the dynamic range shrinks the more electronic units are included in the reproduction chain.

165

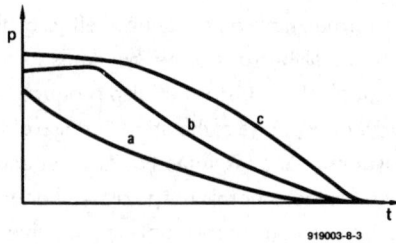

919003-8-3

Figure 3.3.8. Fall-off in sound level after plucking the relevant guitar string has ceased;
(a) not amplified; (b) amplified by overdriven valve amplifier;
(c) amplified by overloaded solid-state amplifier.

When the output of an overdriven valve amplifier drops, the output signal gradually changes from distorted to undistorted, but it does so abruptly in the case of a transistor amplifier. Owing to the overdriving of the amplifier, the sound level of a tone when a string on the guitar is plucked remains virtually constant during the plucking action, but drops off gradually (is sustained for a little while) when plucking ceases. The higher the output voltage of the guitar pick, the more the amplifier can be overdriven. The differences in performance of an overloaded valve amplifier and an overdriven transistor amplifier are shown graphically in Figure 3.3.8.

Guitar amplifiers may be wholly valved, wholly transistorized or hybrid (partly valved and partly transistorized). The simplified block diagram of a typical guitar amplifier is shown in Figure

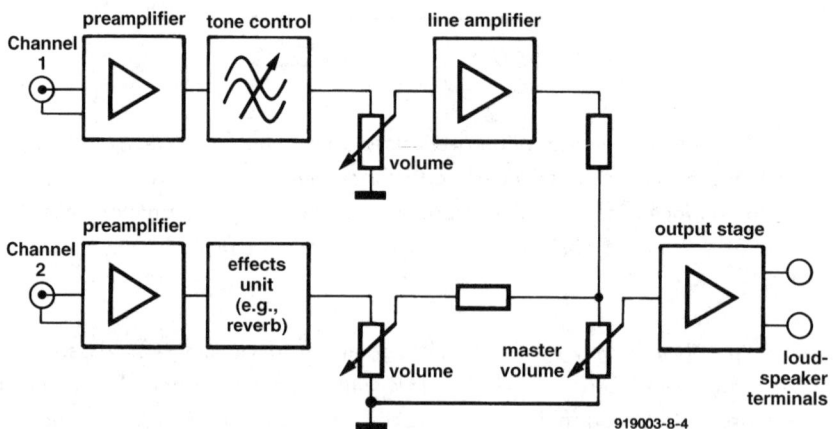

919003-8-4

Figure 3.3.9. Simplified block diagram of a typical guitar amplifier.

3.3.9. Most amplifiers have two input channels which usually produce a basic sound. In the illustration, channel 1 is intended to produce a distorted sound, but channel 2, a clean sound. Each channel has its own, independent volume control, but the overall volume is set with the master volume control.

The volume control in channel 1 enables a distorted sound to be produced at low volume settings: the drive to the amplifier is adjusted until the line amplifier is overloaded. This presupposes that the volume control and that on the guitar are both fully open. The actual output level is then set as desired with the master volume control. The result is that the output amplifier, which is not overdriven, raises a signal that has already become distorted.

Figure 3.3.10 shows the simplified block diagram of a hybrid guitar amplifier, while Figure 3.3.11 shows the inside view of this amplifier from the opened rear.

The preamplifiers are completely transistorized, but the output amplifier is valved. The drive input is connected to a special input stage based on a field-effect transistor (FET). The transfer characteristic of this type of transistor resembles that of a thermionic valve. Consequently, the distortion it introduces looks like that of a valve. This means that this input enables a valve sound to be produced. The output of the input stage is applied to the preamplifier/tone control stage via a potentiometer, which enables the volume of the channel to be set. Coupled to the poten-

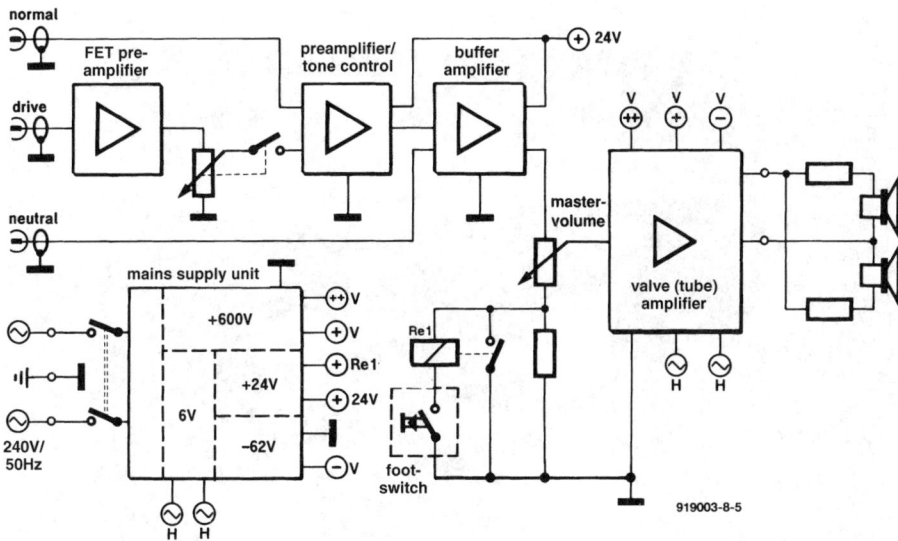

Figure 3.3.10. Simplified block diagram of a hybrid guitar amplifier with three input channels. Each channel has a different number of amplifier stages, so that the input sensitivities of the channels are dissimilar. The footswitch enables selection of one of two volume settings without this affecting the overdrive, and consequently the timbre.

167

tiometer is a switch that enables this input stage to be taken out of circuit. The other input applied to the preamplifier/tone control stage is the normal input.

As mentioned earlier, the drive input enables a valve sound to be produced, while the neutral input is applied to a buffer amplifier that makes the signal suitable for processing by the valved output amplifier. In this signal path there is no possibility of influencing the timbre of the output signal. Moreover, the amplification in this channel is not particularly high, so that it may be used for other instruments, such as a keyboard. It is, of course, also possible to to use a separate guitar preamplifier, whose output must applied to the neutral input, so that the amplifier can provide power amplification only.

The master volume control is linked to earth via a resistor that can be shored by a footswitch-operated relay. When the relay contacts are open, the potential at the wiper of the control, and thus the sound level, is higher than when the contacts are closed. In this way, it possible to use two different sound levels without the necessity of turning a control. It enables, for instance, a particular solo to be emphasized.

The transfer function of the master volume control is shown in Figure 3.3.12. Since the sound level is changed at the master volume control, the sound level changes but not the overall sound picture. A comparable application of a footswitch is shown in Figure 3.3.13. The two output

Figure 3.3.11. Inside view taken from the opened rear of an active hybrid amplifier. Below the two loudspeakers is the three-valve output stage.

168

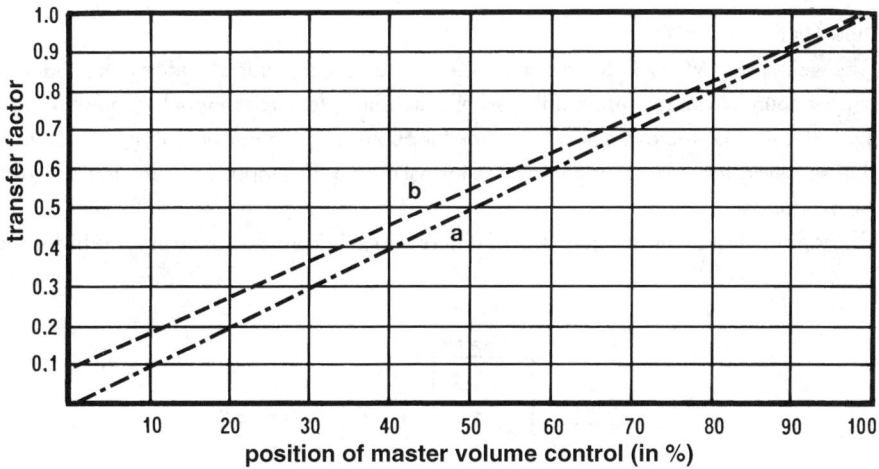

919003-8-7

Figure 3.3.12. When the relay in Figure 3.3.10 is operated, the
characteristic of set sound level in (a) changes to that in (b).
Both depend on the position of the master volume control.

leads of the footswitch are linked to two different input channels of the amplifier.
This enables rapid switching between two sound levels or between two output sounds without
the need of interrupting the performance. The wiring diagram of the footswitch in Figure 3.3.13

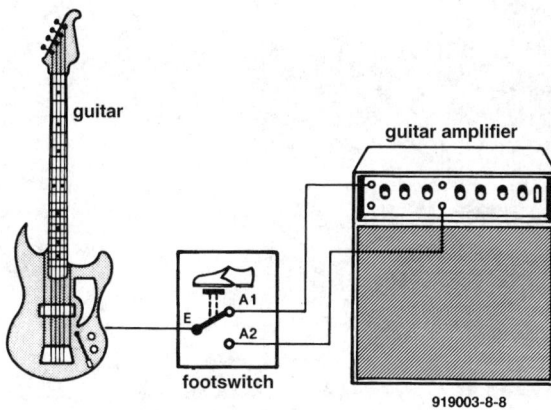

919003-8-8

Figure 3.3.13. A (change-over) footswitch enables one of two
different volumes to be selected

169

is shown in Figure 3.3.14.

The well-known VOX AC-30 in Figure 3.3.15 is a hybrid guitar amplifier that produces a distinctive sound. It has two inputs with dissimilar sensitivity for each channel as shown in Figure 3.3.16. It is thus possible to select the correct input for a given instrument: the one with the higher sensitivity for the instrument with the lowest output. This input is the one that can readily be overloaded.

The tone control circuit in Figure 3.3.17 differs basically from that found in some hi-fi ampli-

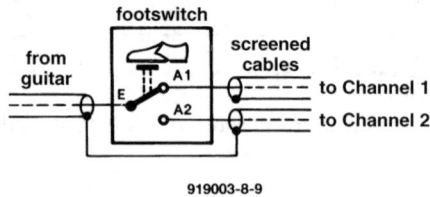

919003-8-9

Figure 3.3.14. Wiring diagram of the footswitch in Figure 3.3.13.
The screen of only one of the two connecting cables should be connected
to that of the guitar lead to prevent earth loops ensuing.

919003-8-10

Figure 3.3.15. The well-known AC-30 from VOX is a hybrid amplifier with a power output
of 30 W. The holes in the front panel indicate that this model has been modified.

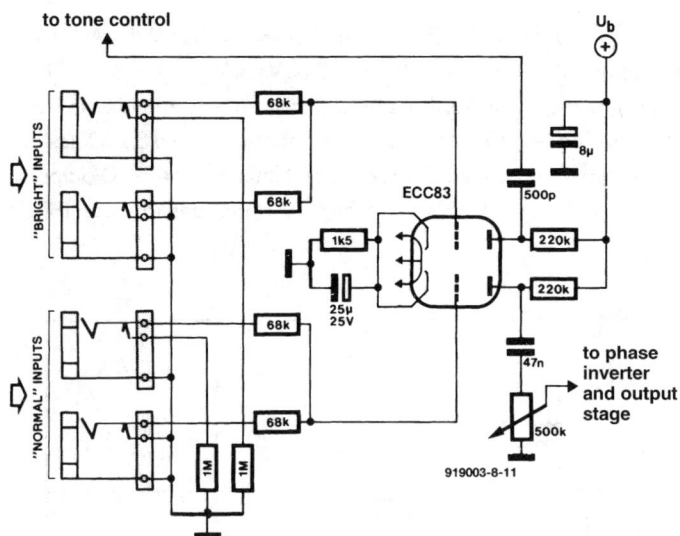

Figure 3.3.16. In the VOX AC-30 each channel has two inputs with dissimilar sensitivity.

fiers or mixers. In contrast to hi-fi circuits, which should produce a straight characteristic when the tone controls are in mid-position, the controls in a guitar amplifier cannot produce a straight characteristic; in any case, no guitarist would want this.

The output stage of the AC-30 is shown in Figure 3.3.18. It uses four Type EL84 valves, dri-

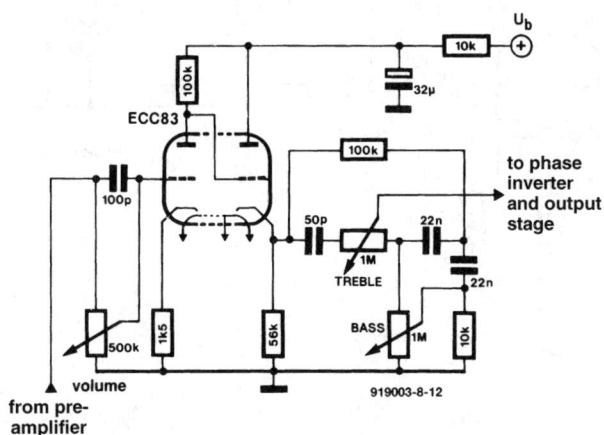

Figure 3.3.17. Tone control circuit of the VOX AC-30.

ven by a Type ECC83 double triode. The specified output power of the AC-30 is 30 W. In an emergency, the output stage can work with two EL84s: V_1 or V_2 with V_3 or V_4. The output power is then considerably lower, of course. The output transformer is intended to work into an 8 Ω or 16 Ω loudspeaker. The AC-30 is equipped as standard with two 8 Ω Celestion loudspeakers, which are intended to be connected in series and linked to the 16 Ω output terminals.

Figure 3.3.19 shows the circuit diagram of the input stages and tone control of the Marshall

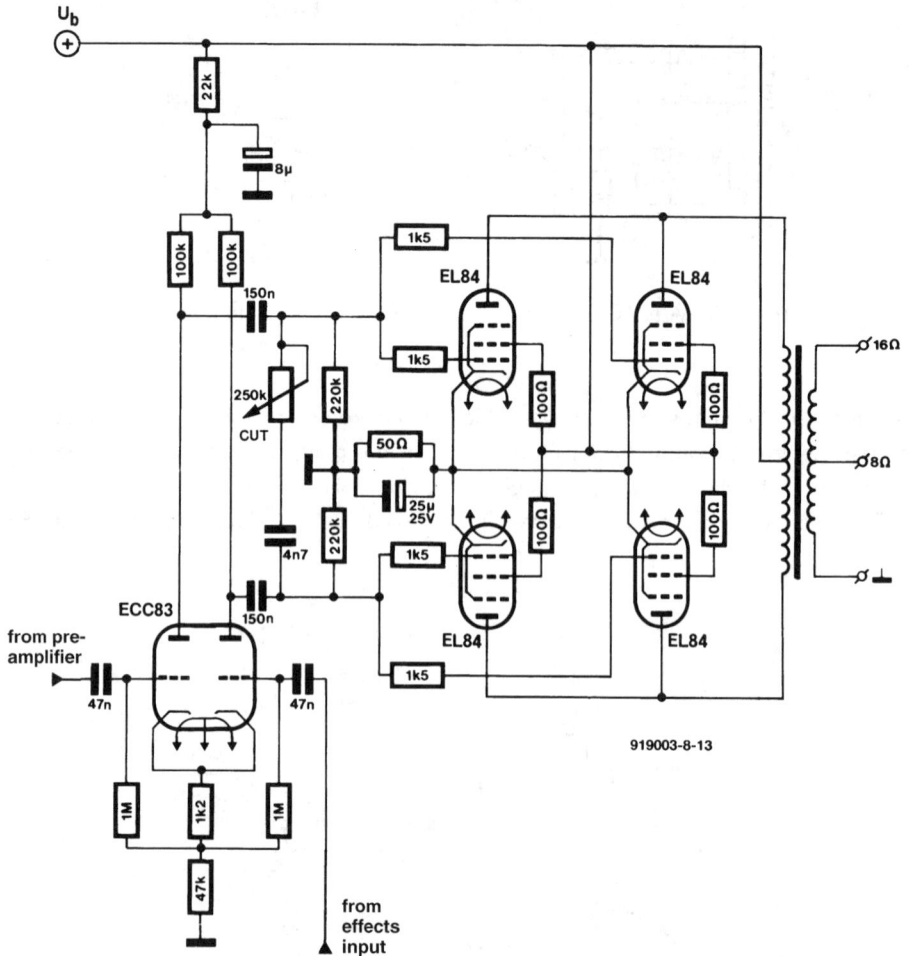

Figure 3.3.18. The output stage of the VOX AC-30 uses four Type EL84 valves in a double parallel configuration. In an emergency, the unit can work with only two of these valves, but the power output is then considerably lower and the valves are overloaded.

JCM800 Model 2204. This is a discrete amplifier which needs separate loudspeakers. Its output power is specified by the manufacturer as 50 W. The 2204 has only one input channel which is, however, provided with two inputs with different sensitivity. Input 'low' has the lower input sensitivity, while an additional amplifier is provided in the signal path linked to input 'high' . The amplifier stages based on V_{1a} and V_{1b} are capable of overloading the stage based on V_{2a}. Preceding the master volume control is a typical three-stage tone control.

Figure 3.3.20 shows the front and rear panels of a Deluxe Reverb II guitar amplifier from Fender. This unit has two channels, each with one input. There are separate volume and bass and treble controls in channel 1.

Channel 2 has more facilities: volume control for the channel, gain of the input stage, and overall master volume control. There is a three-stage tone control: bass, mid-frequencies, and treble. There is also a control for reverb and presence.

The rear panel carries the various inputs and outputs: socket for FOOTSWITCH, PREAMP OUTPUT, to which another or additional power amplifier may be connected. Another or additional preamplifier may be connected to the POWER AMP INPUT socket. The two switches marked GROUNDSWITCH enable the mains earth to be connected at various positions in the amplifier or to be disconnected (not advisable in the United Kingdom – see 3.2.7).

The line output for connection to the mixer is situated beside the loudspeaker output SPEAKER (20 WATTS RMS). The sockets marked REVERB INPUT/OUTPUT enable an echo or other effects unit to be linked to the 2204. At the extreme right is a preset potentiometer for suppressing any mains hum (see 3.3.3).

Apart from the circuits discussed so far, the power soak circuit should be mentioned. Many

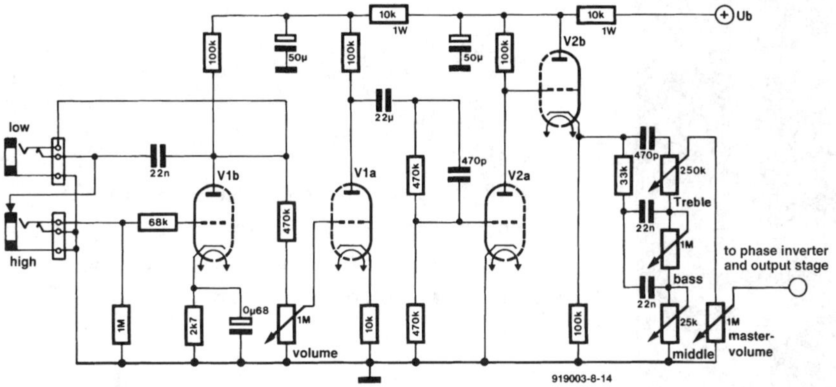

Figure 3.3.19. Diagram of the input circuit of a Marshall 2204.
The dissimilar input sensitivities are obtained by an extra amplifier
stage in the path of the input with the higher sensitivity.

guitar amplifiers provide the best distorted sound if not only the preamplifier(s) but also the output stage(s) are over-loaded. This can perforce only be done at very high volume settings since the amplifier must provide maximum power output. A power soak circuit (see Figure 3.3.21) enables this to be done without damaged eardrums: it converts a large part of the power output into heat so that not much power reaches the loudspeaker(s). The effect on the timbre is negligible. Figure 3.3.22 shows how the power soak circuit is inserted into the sound production chain.

The resistors in a power soak circuit must clearly be power types. Also, it is important that the amplifier is terminated cor-

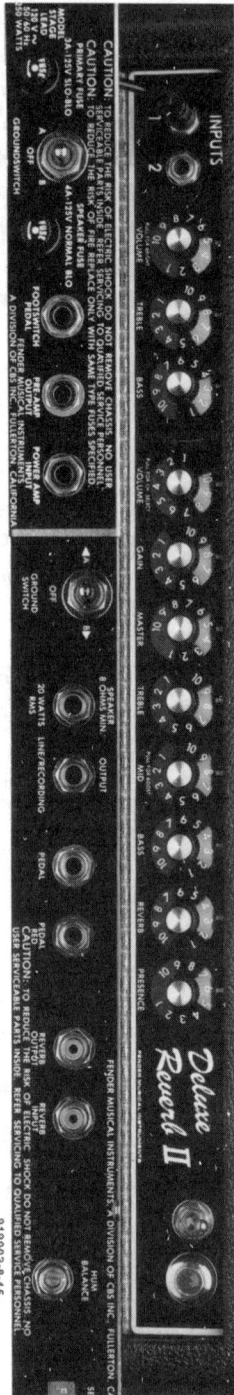

Figure 3.3.21. Diagram of a basic power soak circuit.

rectly at all times. The value of resistors R_1 and R_2 in Figure 3.3.21 is calculated from:

$$R_1 = R_{LS} \left(1 - \sqrt{P_{LS}/P_a}\right)$$

$$R_2 = R_{LS} / \left(\sqrt{P_a/P_{LS}} - 1\right)$$

where P_{LS} is the wanted power to the loudspeaker(s), P_a is the nominal power output of the amplifier, and R_{LS} is the loudspeaker impedance (normally 4 Ω or 8 Ω). The power dissipated in the resistors is calculated with Eq.10:

$$P_{R1} = R_1 I_a^2$$

Figure 3.3.20. Front and rear panels of the Deluxe Reverb II guitar amplifier from Fender.

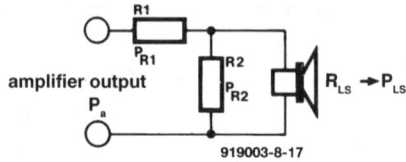

174

$$P_{R2} = R_2 I_{R2}{}^2$$

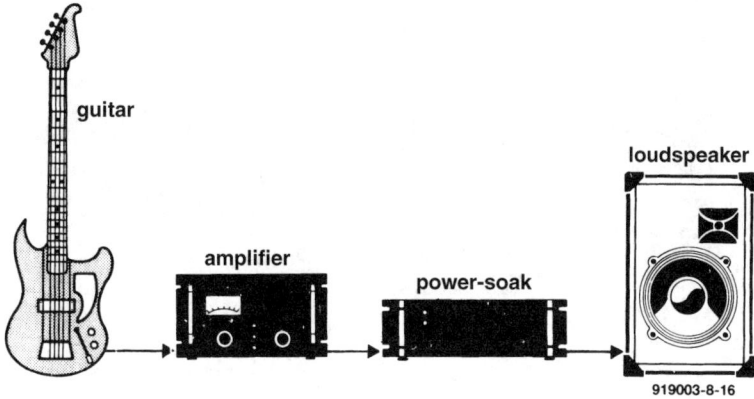

Figure 3.3.22. The correct position of a power soak circuit
in the production chain of an electric guitar.

Figure 3.3.24 shows some DIY circuits for a power soak that can be used in many cases. The resistors should all be sealed in a high insulation cement or ceramic box as shown in Figure 3.3.23. Bear in mind that the resistor values in the circuit diagram are minimum values. The power soak should be fitted in a sturdy metal case with a number of ventilation slots or round openings. The circuit is best constructed on a ceramic prototyping board.

The power soak should have plug-and-socket facilities so that it can be added to, or removed from, the signal path to make full power available again. As mentioned earlier, a valve ampli-

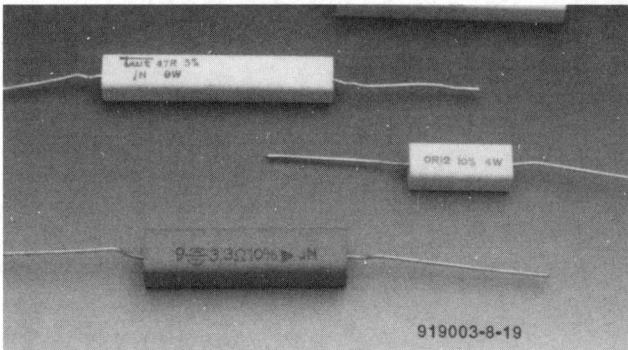

Figure 3.3.23. Power resistors sealed in a high insulation cement or ceramic box.

fier must not be used without load. So, before the amplifier is switched on, it should always be ascertained whether the loudspeaker(s) with or without power soak are properly connected to the amplifier output.

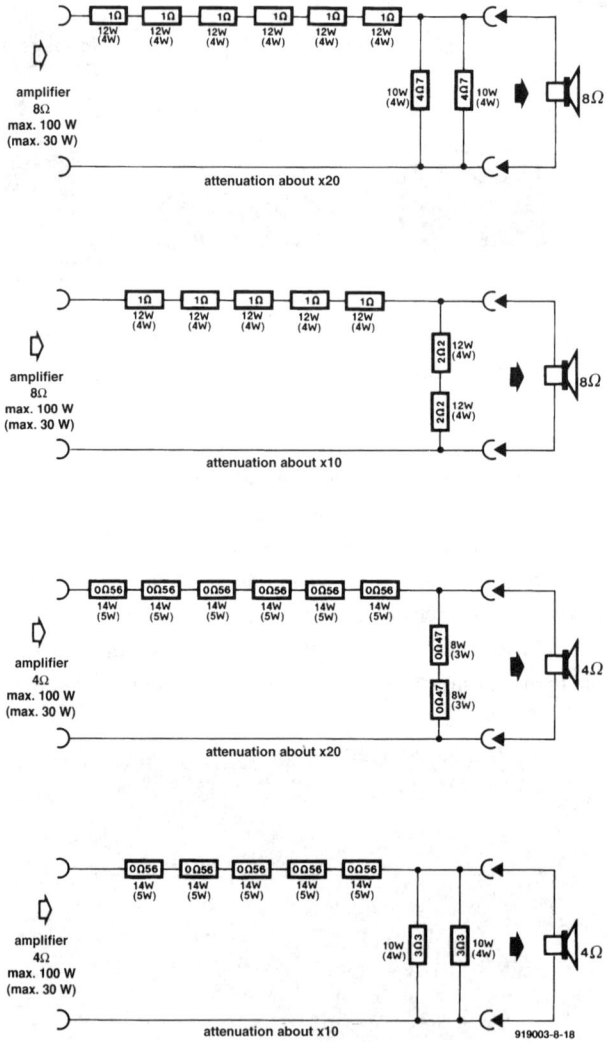

Figure 3.3.24. Power soak circuits for amplifiers with an output impedance of 4 Ω or 8 Ω, all computed for an attenuation of ×10 or ×20.

3.3.3 Bass guitar amplifiers

A bass guitar amplifier must have a considerably higher power output than a standard guitar amplifier to enable it to emulate or surpass the sound level of the percussion section. The reason for this is that for equal sound levels the requisite power increases with decreasing frequency. Usable powers lie between 80 W and 150 W (measured as described in 3.1). In respect of output power, bear in mind that doubling it is only just audible (see 1.2 and 1.3).

The bass guitar amplifier is co-responsible for the produced sound. In days past, valve amplifiers with a curved frequency characteristic were used. In those early units, the possibilities for influencing the final sound were modest, although operation of the unit was simple. A bass guitar amplifier then was judged primarily on the final sound.

Although similar amplifiers are still available, the modern trend is toward linear amplifiers, which raise the sound but do not influencece it. These amplifiers have an extensive tone control which enables virtually any sound to be obtained by the introduction of linear distortion, that is, a modification of the frequency response. Non-linear distortion in a bass guitar amplifier is not desired.

A drawback of these modern amplifiers is their complicated operation: a musician without a technical background will seldom be able to use all the facilities provided properly. Normally, he/she will simply set up the controls by ear.

Bass guitar amplifiers are also available with valves, transistors, or a mixture of these. Nowadays, valves are normally used only in amplifiers with a unique eigensound, that is, in which the final sound is determined entirely by the valves. The block diagram of such an ampli-

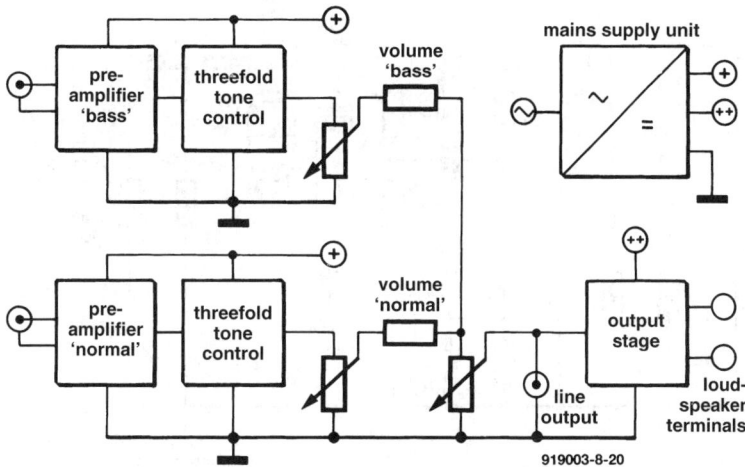

Figure 3.3.25. Simplified block diagram of a typical bass guitar amplifier.

177

fier is shown in Figure 3.3.25. Linear amplifiers with extensive tone control are invariably solid-state designs. Additionally, there are hybrid amplifiers on the market that use valves as well as transistors and/or integrated circuits (ICs). Like standard guitar amplifiers, bass guitar amplifiers are available as discrete units or integrated with loudspeaker(s) and equalizer as a combo. The latter normally have a presence control, overall distortion switch, and level control plus sockets for external effects units.

The influence of a bass guitar amplifier on the produced sound is much smaller than that of a standard guitar. The linear distortion produced by it may be imitated by a large mixer with extensive tone control facilities. This is why many bass guitarists normally have no objection to play straight into the mixer (in contrast to standard-guitar players). This may be of importance if the bass guitar amplifier has a line or DI output (see Section 6.3). Many amplifiers have, apart from the usual 6.3 mm audio sockets, a balanced input, which is is, of course, beneficial when noise and hum have to be combated.

The circuit of the input stages of the Bassman 135 bass guitar amplifier is shown in Figure 3.3.26, while a general view of this amplifier is given in Figure 3.3.27. This amplifier has two input channels, the first of which is intended for electric bass guitars. The second channel can accommodate other instruments, such as a keyboard.

The channel shown in Figure 3.3.26 is intended for bass guitars. The input sensitivity is 162 mV and the input resistance is 136 kΩ. The input sensitivity of the other channel is 81 mV and the input resistance is 1 MΩ. The figures in both cases apply to an output power of 135 W.

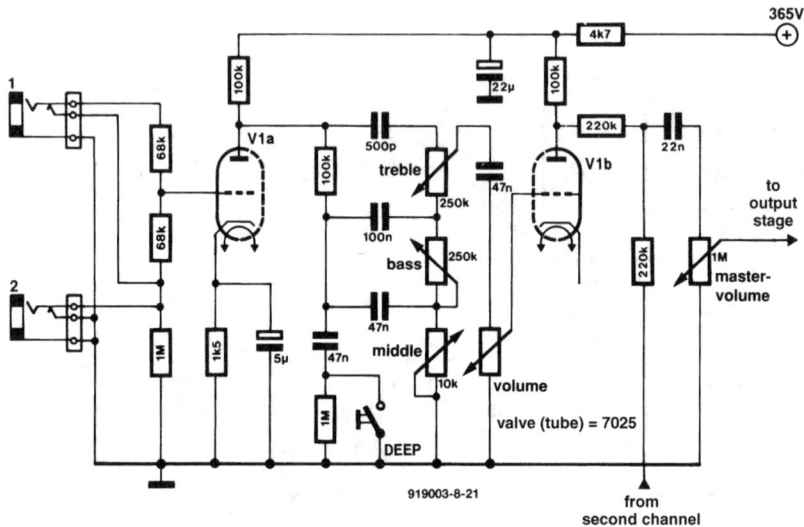

Figure 3.3.26. Input stages of the Bassman 135 from Fender.

178

Figure 3.3.27. General view of the Bassman 135 from Fender.

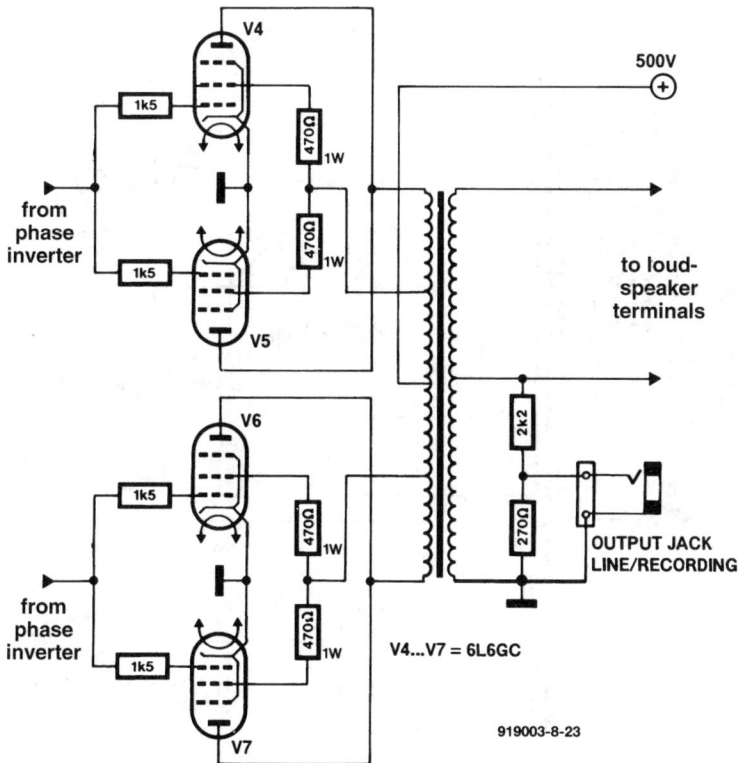

Figure 3.3.28. Circuit diagram of the output stages of the Bassman 135.

The tone control, following the first valve stage, V_{1a}, is typical of musicians' amplifiers and offers separate control of bass, mid-range, an treble frequencies. The switch marked DEEP enables further influence of the sound. The tone control is followed by the volume control and this in turn by the second valve stage, V_{1b}.

The output stages of the amplifier are shown in Figure 3.3.28. They comprise four output valves, V_4–V_7, which are connected in a double parallel arrangement to obtain the nominal power output of 135 W.

The circuit of the power supply for the amplifier is shown in Figure 3.3.29. Noteworthy is the NUM BALANCE control, which is for setting a symmetrical heater voltage. Valves have a heater whose temperature is raised to very high levels by the application of a 6.3 V a.c. potential. This may cause mains hum, which can be reduced to a minimum by the NUM BALANCE control. Normally, this is preset in the factory, but when a valve has been replaced, it may be necess-

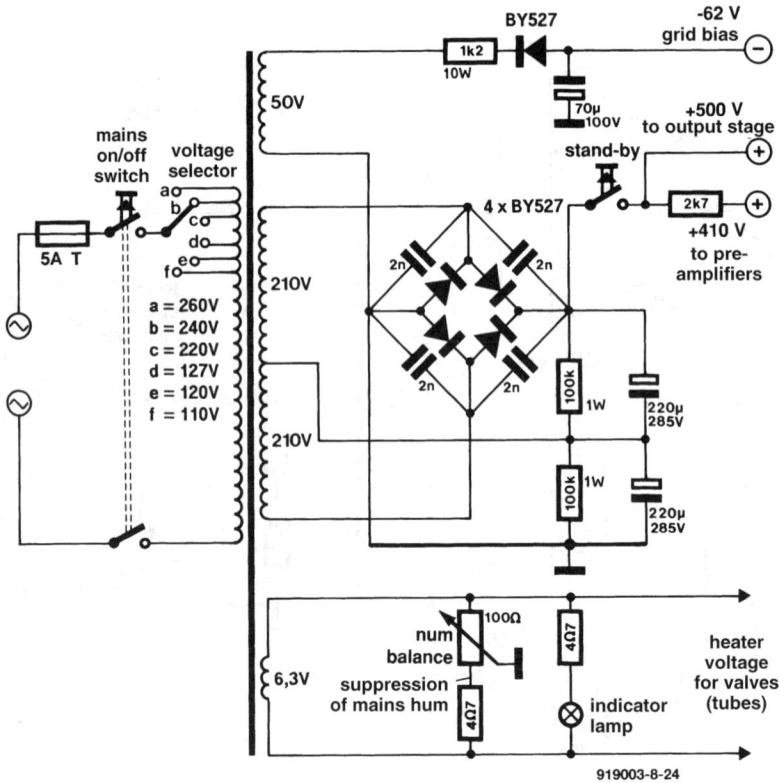

Figure 3.3.29. Power supply of the Bassman 135.

180

Figure 3.3.30. Front panel of the solid-state bass guitar amplifier Type DP300 Bass from Laney.

ary to readjust the control slightly.

The STANDBY switch enables the anode voltage of the valves to be switched off. The heaters remain powered, however, so that the amplifier is immediately operational when the anode voltage is switched back on. It is advisable to switch off the anode voltages during intervals and interruptions, since this lengthens the life of the valves; it also prevents hum, crackles and plops that may occur when additional instruments are linked in. The mains on/off and standby switches are frequently fitted at the rear of the amplifier.

Figure 3.3.30 is an illustration of the front panel of a Type DP300 bass guitar amplifier from Laney. This is a solid-state amplifier with very extensive tone control facilities. The knob at the extreme left serves to set the amplification factor of the input stages. At the centre is a 9-band equalizer (see Chapter 5), which is, in effect, a tone control that can adapt the sound to almost any requirement. When all nine slide potentiometers are in their mid-position, the amplification is linear. An additional slide potentiometer enables influencing the operation of the equalizer/tone control. When this potentiometer is fully down, the discrete frequency bands of the equalizer can only be amplified. When it is fully up, the bands can be attenuated only.

When it is in the mid-position, the equalizer works in the same way as that of a hi-fi system: each and every frequency band can be amplified or attenuated independently as required. If needed, the equalizer can be by-passed by a separate switch. The large knob at the right is the master volume control. Another control is the noise gate, with which the input stages are enabled only when the bass guitar is actually being played. During intervals and interruptions, the input stages are automatically short-circuited so that there is not any noise or hum audible. There is also a balanced XLR input.

3.3.4 Keyboard (instrument) amplifiers

In days gone by, the keyboard amplifier had a role to play in shaping the output sound. This is no longer true of modern keyboard amplifiers, which, like a hi-fi amplifier, operate linearly, that is, they do not influence the shape of the signal. This is the reason why modern keyboard amplifiers are fully transistorized. In some respects, their circuit resembles that of a hi-fi amplifier. However, the keyboard amplifier must meet certain specific requirements. First, pop and rock groups seldom use just one keyboard, so that the amplifier must have a sufficient number of input channels to enable several keyboards to be used simultaneously. Also, a substantial power output is required, since the frequency response goes down to the lowest frequencies.

The guideline in choosing an amplifier must be that the sound level of the percussion sec-

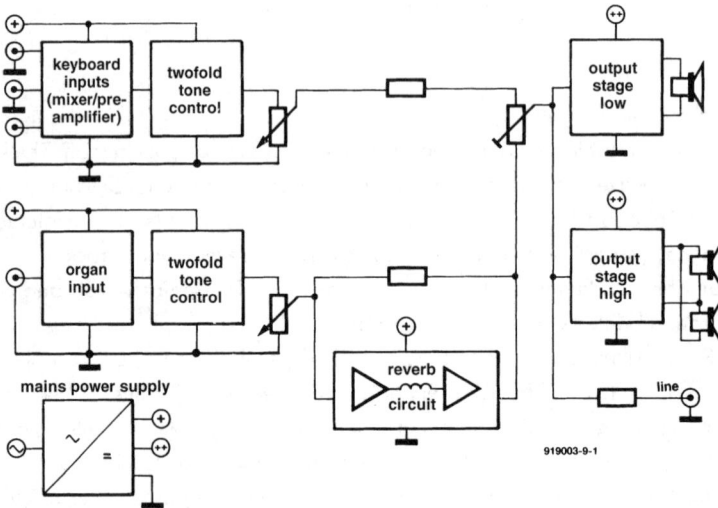

Figure 3.3.31. Simplified block diagram of a keyboard cum organ amplifier with separate output stages for bass and treble reproduction.

182

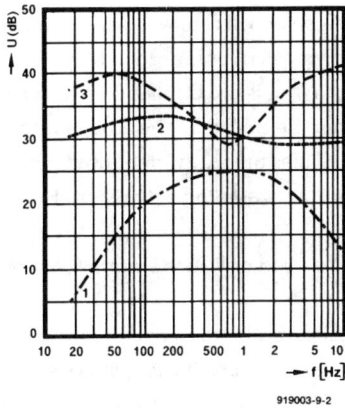

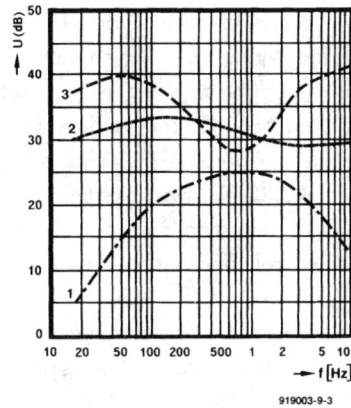

Figure 3.3.32. Typical frequency vs output characteristics of a keyboard cum organ amplifier for various positions of the tone controls. The characteristics at the left pertain to the
organ input: bass and treble fully anticlockwise (1), in mid-position (2), and fully clockwise
(3). The curves at the right refer to the keyboard input: controls fully anticlockwise (1), in
mid-position (2), and fully clockwise (3).

tion must be equalled or surpassed. Taking into consideration the efficiency of modern loudspeakers, this means that the output power must lie between 80 W and 150 W.

It is, of course, possible to assemble a keyboard system from discrete units: mixer, power
amplifier and loudspeaker(s). With the rich variety of units on the market, this should not pose
any problems, as long as the mixer and power amplifier have a sufficient number of input channels (each keyboard requires one). The input channels should preferably be of the dual type,
since many keyboards provide a two-channel output (which is not stereo!).

An assembly of discrete units has drawbacks: operation is more complicated and the cabling
may cause nightmares. Even rack-mounting (for which 19-in units are needed) does not
improve this situation. Some manufacturers therefore market special keyboard amplifiers with
built-in mixer or with integral loudspeaker(s). Most small power mixers may be used as keyboard amplifier, but to reproduce the highest and lowest frequencies of the keyboard several
loudspeakers are needed. A single wide-range loudspeaker does not perform as well as a combination of woofer, mid-range driver, and tweeter (see Chapter 4).

Except for electronic organs with rotating loudspeakers (which influence the sound), keyboards can be recorded directly, that is, without microphone if a public-address installation is
used. For this, it is necessary for the amplifier or mixer to have a line, or better, DI, output (see
Section 6.3). The difference between a line output and a DI output is that the latter is electrically isolated from the amplifier.

The simplified block diagram of a keyboard amplifier is shown in Figure 3.3.31. The amplifier has four inputs, one of which is intended for connecting an electronic organ; the other three are for keyboards (synthesizers, and so on). The keyboard inputs are applied to a mixer/preamplifier so that differences in level between the three can be nullified. The tone controls for organ and keyboard are also separate, although their facilities are exactly the same as shown in Figure 3.3.32. The two potentiometers enable the separate presetting of the output levels of the organ and keyboard inputs.

The integral effects circuit, here a reverb circuit, can be used only in the signal path of the organ.

The two channels are combined by a common preset potentiometer and then applied to the output stages. Note that here is no question of dual channel reproduction, but of separate amplification of the bass and treble frequencies. The transfer function of the two output stages is given in Figure 3.3.33.

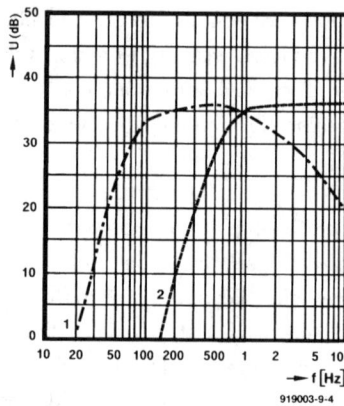

Figure 3.3.33. Frequency response characteristic of the two output stages in Figure 3.3.31: (1) high-frequency amplifier; (2) low-frequency amplifier.

The circuit diagram of a typical organ input stage, here the Type DC300E keyboard and organ amplifier from Dynacord), is shown in Figure 3.3.34. A photograph showing the controls for the organ input is shown in Figure 3.3.35. A level of 10 mV at the lower (more sensitive) input (OH) is sufficient to drive the amplifier to full output. The circuit diagram of the mixer amplifier of the same equipment is shown in Figure 3.3.36. Three keyboards can be connected to this amplifier. For full output of the amplifier, a level of 30 mV is needed at one or more of the inputs. The amplification factor of each of the amplifier stages is $A = 2.7$. Potentiometers R_{118}–R_{120} enable the individual setting of the sound level in each of the channels.

The output level of many older keyboards is very low, so that the control range of a modern keyboard amplifier is not sufficient to compensate the differences in level between different keyboards. In principle, it would be possible to attenuate the signal from the keyboard with the highest output via a potential divider, but owing to the lowered signal voltage, the signal-to-noise ratio would degrade and the risk of hum and noise would increase.

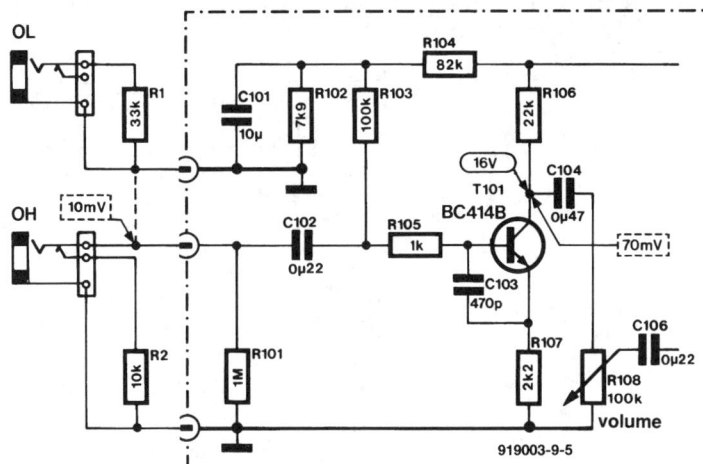

Figure 3.3.34. Circuit of the organ input stages of transistorized keyboard and organ amplifier Type DC300E from Dynacord. The two inputs have a dissimilar input sensitivity: that of OH (high) is greater than that of OL (low).

Figure 3.3.35. Organ input section of the DC300E from Dynacord.

It is, therefore, better to amplify the signal with a small amplifier such as that shown in Figure 3.3.37. This circuit is readily built on a small piece of prototyping board. The supply voltage, U_b, may be derived from the supply lines of the keyboard. If this does not have a 15 V line, the value of resistor R_1 must be changed according to Table 3.3.1.Since the amplifier is very small, it may well be possible to build it into the keyboard as shown in Figure 3.3.38. The operation of the amplifier is shown in the oscilloscope trace in Figure 3.3.39, and the transfer characteristic in Figure 3.3.40. The amplification factor is 2.5. If a higher values is needed, the value

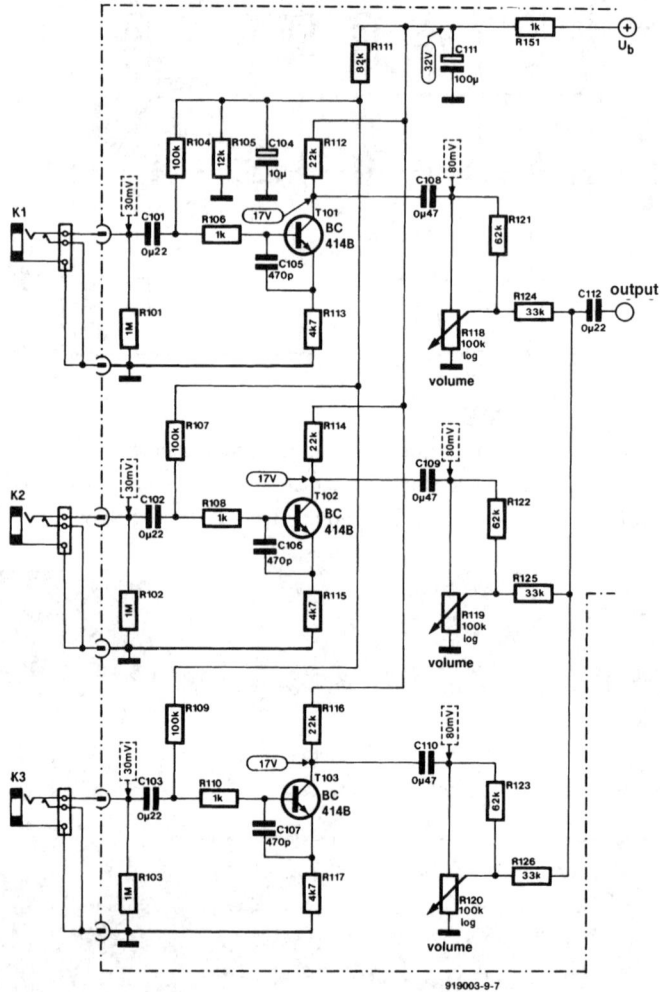

Figure 3.3.36. Circuit diagram of the keyboard inputs of the DC300E.

U_b	R_1
12 V	wire bridge
15 V	1.8 kΩ
18 V	3.3 kΩ
21 V	4.7 kΩ
24 V	6.8 kΩ
33 V	12 kΩ
40 V	18 kΩ

Table 3.3.1

of R_6 may be lowered. This should be done with some care, however, to avoid the amplifier being overloaded, which leads to serious distortion of the signal.

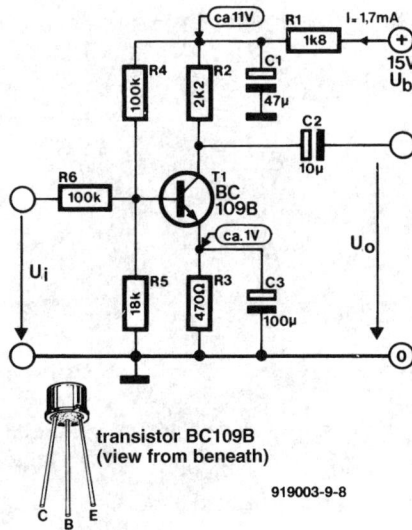

Figure 3.3.37. Circuit diagram of a small amplifier to raise the output level of an older type of keyboard.

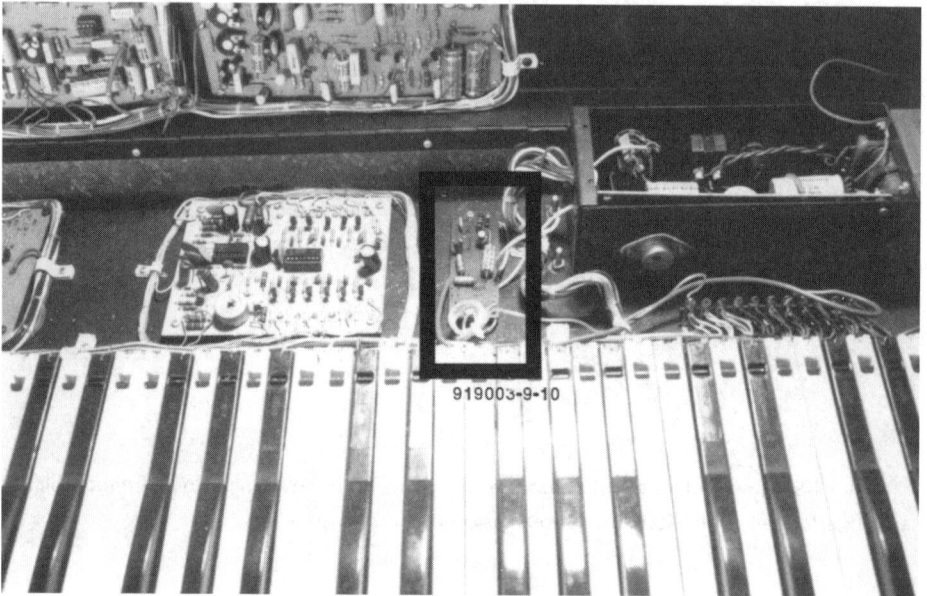

Figure 3.3.38. How the small amplifier may be fitted in an older type keyboard.

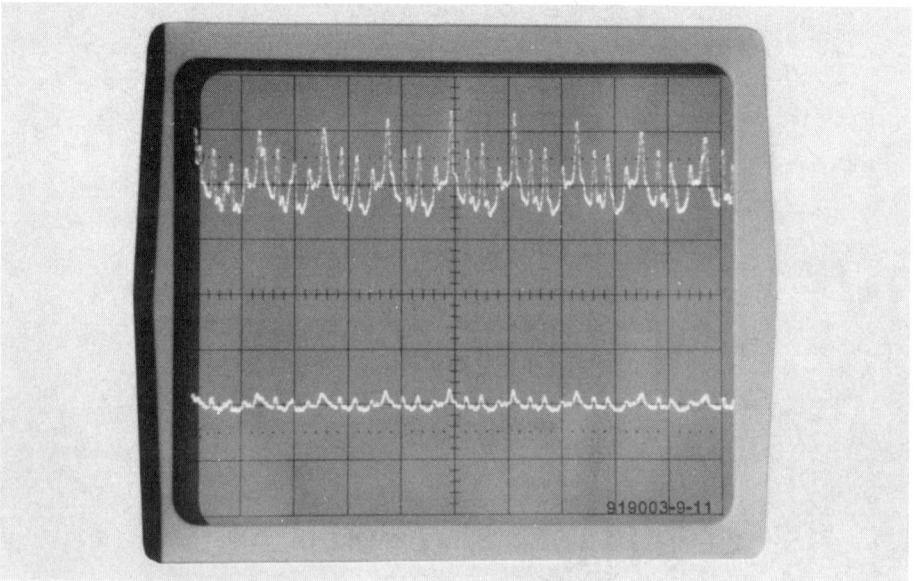

Figure 3.3.39. Input voltage (lower trace) and amplified signal (upper trace)
of the small amplifier in Figure 3.3.37.

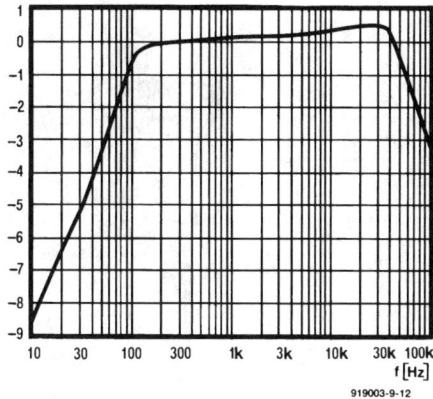

Figure 3.3.40. Transfer characteristic of the amplifier in Figure 3.3.37.

3.3.5 Some facts about thermionic valves

Today musicians' amplifiers, special hi-fi amplifiers and transmitters are the only pieces of equipment that may use thermionic valves (electron tubes). In most other equipment, transistors and other semiconductors, as well as integrated circuits (ICs), have taken their place. The trend to using solid-state amplifiers is also noticeable in the musicians' world, except in guitar (lead as well as rhythm) where the valve still reigns supreme. Apart from new models, there are still countless older valve guitar amplifiers in use all over the world.

A thermionic valve is intended to process very small alternating voltages. It consists of an evacuated glass or metal cylindrical tube fused onto a base. The base is made of pressed glass through which the terminals pins protrude. Inside the tube, a system of electrodes is mounted on these pins. The simplest amplifier valve is one with three electrodes, and is called triode (from the Greek tri = three). These electrodes are called cathode (k), anode (a), and control grid (g_1). The circuit symbol and the internal construction of a triode are shown in Figure 3.3.41.

The cathode is brought to its operating temperature by the heater around which it is placed. The control grid normally consists of a spiral of thin wire. Since the anode can get fairly hot during operation, it is normally a metal cylinder with a relatively large surface area. In preamplifier valves, two such triodes are often placed in the same glass tube: these are called double triodes. Valves for output stages normally contain more than the control grid: the screen grid (g_2) and the suppressor grid (g_3). Valves with a control grid and a screen grid are called tetrodes, and those with a suppressor grid in addition are called pentodes.

When the cathode is heated and reaches its normal operating temperature, electrons escape from its surface and form a cloud, called the space charge, around the cathode. When

189

triode

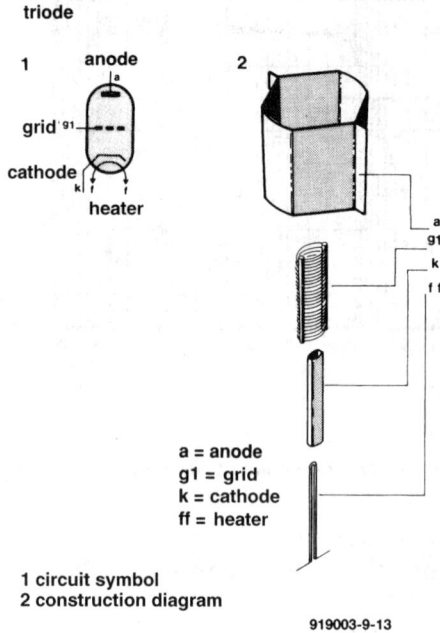

1 circuit symbol
2 construction diagram

919003-9-13

a = anode
g1 = grid
k = cathode
ff = heater

Figure 3.3.41. Circuit symbol and construction of a triode.

a direct voltage is applied between the cathode (negative) and the anode (positive), electronics around the cathode will be attracted by the anode. The resulting stream of electrons has a constant magnitude that depends on the anode voltage, and is called anode current. The construction of the control grid allows electrons to pass through it. A potential applied between the cathode and control grid influences the anode current. If this potential is negative, the electron stream is slowed down since the grid then repels the electrons. This repulsion becomes stronger as the potential becomes more negative. The anode current then becomes smaller. A positive grid potential accelerates the attraction by the anode which speeds up the electrons and the anode current becomes larger. In this way, a small change in grid potential causes a much larger change in anode current and therefore in the power produced by the valve.

Since electrons have virtually no mass and therefore hardly any inertia, very little power is needed to accelerate or decelerate them. This means that only a potential and negligible current are required to vary the anode current. Because of this, a thermionic valve can be used to amplify very small signal voltages, even at very high frequencies. The nature of the amplification, that is, the relationship between input signal and output signal and the drive performance of the valve together largely determine the unequalled dynamic performance and typical mellow, rounded sound produced by a guitar amplifier.

In the past, vast numbers of valves of all kinds and for all sorts of application were produced and sold each year. Nowadays, valves for use in audio engineering are produced in much smaller numbers and mainly in countries with low labour costs. There are specially selected high-quality valves on the market at very high prices, which purport to improve the audio sound. However, the sound coming from a loudspeaker is the result of many factors, such as the design and construction of the amplifier and the manner in which the instrument is played. An amplifier that uses certain types of valve does not sound audibly different if these valves are replaced by valves of the same type and specification, but produced by a different manufacturer. The specification of a valve gives its most important parameters, such as the mutual conductance, the open-load amplification, the internal resistance, parasitic capacitances, and the I/U characteristics. Some of these, referring to a Type EL84 output valve, are summarized in Table 3.3.2.

heater voltage	6.3 V
heater current	760 mA
heating	indirect
maximum anode dissipation	12 W
Class B push-pull	
anode voltage	300 V
screen grid voltage	300 V
control grid bias	−14.7 V
anode current	2×7.5 mA
a.f. power output	17 W
Class AB push-pull	
anode voltage	300 V
screen grid voltage	300 V
cathode resistance	130 Ω
anode current	2×36 mA
a.f. power output	17 W

Table 3.3.2. Brief specification of EL84

Since the data have a certain spread owing to unavoidable manufacturing tolerances, the valves in a push-pull stage must always be replaced, if needed, by valves from the same production run. In practice, this means that when new valves are being purchased, it should be ensured that they are from the same manufacturer and were produced at the same time. This ensures that any differences and asymmetry in the output stage is kept to a minimum. The valves in a preamplifier, however, can be replaced, when needed, one by one. A selection of valves for use in preamplifiers and output stages is shown in Table 3.3.3.

Type number	Classification
6CA7	output pentode
6L6GC	output pentode
EL34	output pentode
EL84	output pentode
ECF83	triode pentode
7025	double triode
12AT7	double triode
12AX7	double triode (low microphony)
ECC81	double triode
ECC83	double triode

Table 3.3.3. Some types of valve and their classification.

When to replace valves (although they are not clearly defective) is a vexed question. In general, those in preamplifier stages should last a good four years. When the getter* in output valves is transparent and a clear brown instead of silvery colour, it is advisable to replace the valve(s) to prevent unpleasant surprises during a performance. Reduced amplification and/or incapability of being overloaded often point to faulty valves.

Sometimes microphony may occur when the valve is subjected to mechanical vibrations. The parts of the internal system of electrodes then vibrate in unison, which causes spurious sounds to be produced. This phenomenon points to a worn-out or faulty valve. If a valve is suspected to be faulty, tap with the handle of a small screwdriver against the tube (with the amplifier switched on); a faulty valve normally causes loud crackles from the loudspeaker; a good valve does not.

---

* An alkali metal introduced into the tube during manufacture. It is fired after the tube has been evacuated to react chemically with, and eliminate, any residual gases.

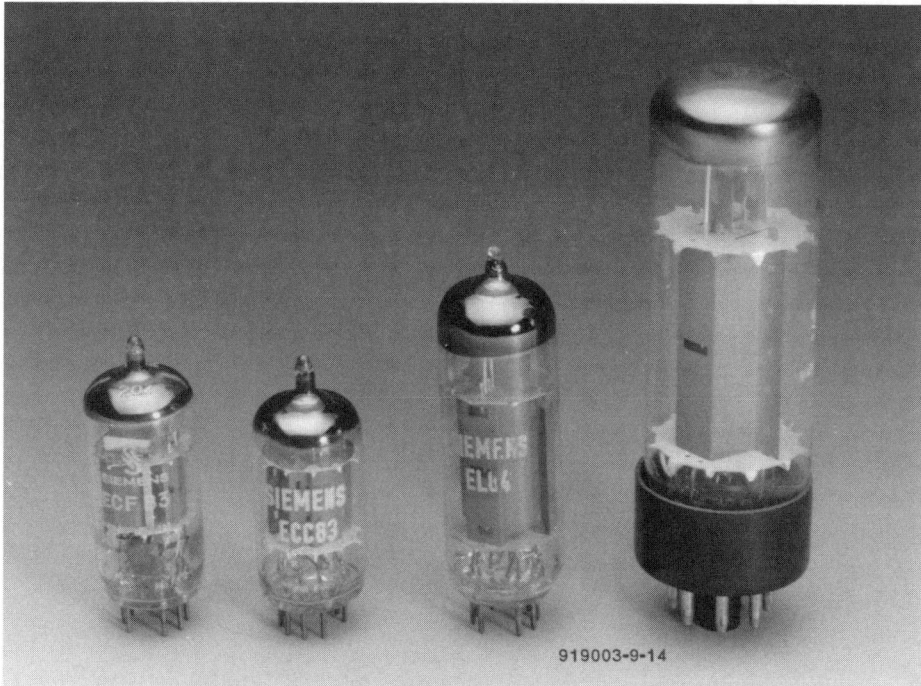

919003-9-14

Figure 3.3.42. Some valves frequently encountered in musicians' amplifiers.
From left to right: ECF83. ECC83, EL84, and EL34.

Although thermionic valves are generally fairly robust, they need to be treated with some care. If at all possible, valve amplifiers should be carried rather than be transported on a trolley to avoid mechanical shock and vibration. Separate loudspeakers also contribute to a longer life of the valves: sound vibrations in an integral amplifier or mixer can be detrimental to valves in the long term. It is good practice to make use of the standby switch during intervals and interruptions in the performance. This switches off the high supply voltage to anodes and screen grids of the valves, but the heaters remain on. Switching the heaters on and off repeatedly causes them to age prematurely.

Finally, the ventilation slots or openings in an amplifier or mixer should never be covered since these are essential for the diffusion of the heat produced inside the unit.

4. Loudspeakers

A loudspeaker serves to convert the electrical energy provided by the power amplifier into acoustic energy (sound). It could be said that a loudspeaker is the exact opposite of a microphone, although, of course, the power levels are quite different. A loudspeaker used in a music hall must be capable of producing a high sound pressure level. The requirements of loudspeakers used by pop and rock groups are quite different from those used in a hi-fi installation. The latter is to reproduce sound as faithfully as possible, but in the former this is only so in the case of public-address and vocal installations. In all other cases, the loudspeaker, like the associated amplifiers, determines the reproduced sound, which means that, apart from a high sound pressure level, it must introduce a certain amount of distortion. Note that a loudspeaker is a combination of one or more drive units, an enclosure (sound board) and, if several drive units are used, a crossover network or filter.

4.1 Introduction

An overview of current types of drive unit (or driver) is shown in Figure 4.1.1. The most important group is formed by dynamic units. Electrostatic drivers are used primarily in hi-fi installations. The principal type of drive unit remains the moving-coil which was patented in 1925 by Rice and Kellog.

The basic construction of a dynamic cone drive unit is shown in Figure 4.1.2. Note that there are a number of variants on this design. The diaphragm, which is made of paper, man-made fibre or light metal, is flexibly suspended at its edges in a frame. At the top of the conical membrane is the voice or speech coil which can move freely in an annular air-gap between the poles and a permanent magnet. A dust cap prevents dust particles from entering the air-gap.

The operation of a loudspeaker is best understood by reference to 1.1.1 in which the principle of voltage generation by induction was discussed. When a low-frequency voltage is applied to the voice coil, a force is exerted on it, which causes it to move in the air-gap between the magnet and poles. The direction into which movement takes place depends on the magnitude and polarity of the alternating voltage. Since the voice coil is fixed to the diaphragm, this follows the movements of the voice coil so that the surrounding air particles are set into vibration.

The most expensive item of a drive unit, particularly low- and mid-frequency types, is the

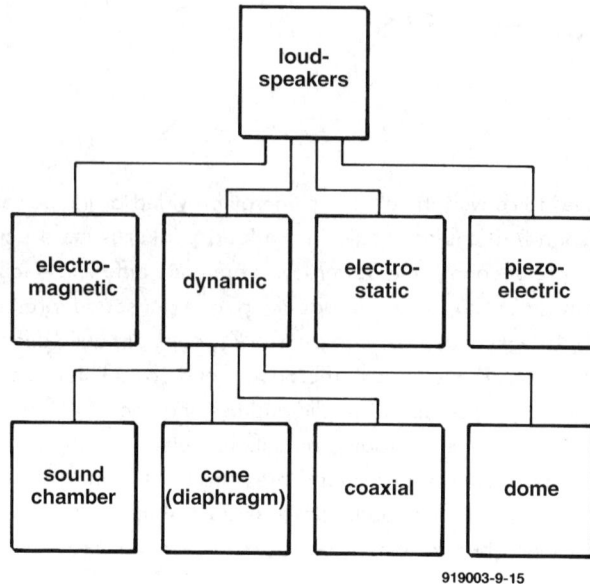

```
                    ┌──────────────┐
                    │    loud-     │
                    │   speakers   │
                    └──────────────┘
```

┌──────────────┐ ┌──────────────┐ ┌──────────────┐ ┌──────────────┐
│ electro-   │ │   │ │ electro-   │ │   piezo-   │
│ magnetic   │ │   dynamic   │ │   static   │ │ electric   │
└──────────────┘ └──────────────┘ └──────────────┘ └──────────────┘

┌──────────────┐ ┌──────────────┐ ┌──────────────┐ ┌──────────────┐
│   sound   │ │   cone   │ │   │ │   │
│   chamber   │ │ (diaphragm) │ │   coaxial   │ │   dome   │
└──────────────┘ └──────────────┘ └──────────────┘ └──────────────┘

919003-9-15

Figure 4.1.1. Current types of loudspeaker.

permanent magnet. The magnetic field in the air-gap must be as strong as possible. Today, magnets may be constructed from ceramic materials. The diaphragm must be as stiff as possible. To keep its cost low, it is usually made of paper or thin cardboard, but more expensive drive units have a membrane made of polystyrene foam or light metals.

The suspension is of great interest—some variants are shown in Figure 4.1.3. Hi-fi drive units normally have a flexible surround. A slightly stiffer surround obtained by pressing creases in the surround enables a better centring of the diaphragm which enhances the reliability at large loads. This is the reason that drive units used in music loudspeakers and those used in discotheque loudspeakers invariably use a rigid surround.

A drive unit forms a vibrating system with a certain resonant frequency, which determines the lower limit of the frequency range of the unit; below this limit there is practically no radiation of sound. The desire for a light and rigid cone in combination with a light, sturdy voice coil leads to irreconcilable requirements. The impulse response and the reproduction of high tones are improved, but the lightly vibrating system so formed has a high resonant frequency, which is a severe handicap for the reproduction of bass frequencies. Only a well-considered compromise can lead to a good-quality drive unit.

A dome radiator is very similar to a cone driver as far as construction and operation are concerned, but it does not have a diaphragm: the sound is radiated entirely by the

196

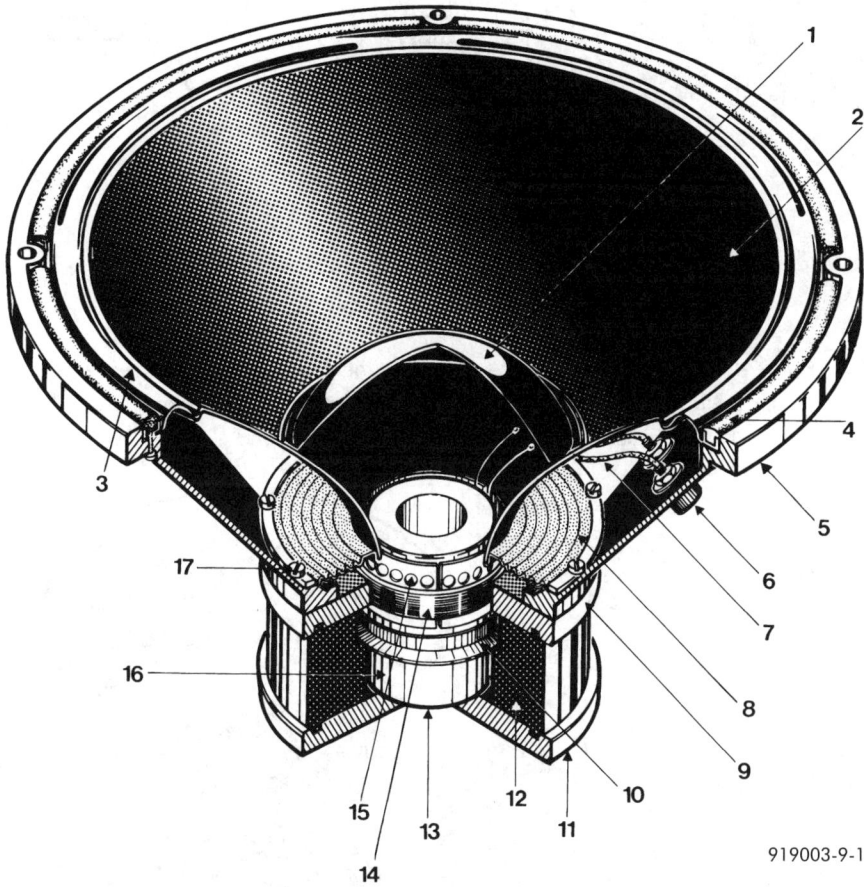

Figure 4.1.2. Cross-section of an electromagnetic drive unit.

919003-9-16

1 = dust cap
2 = diaphragm or cone
3 = suspension
4 = surround
5 = chassis
6 = lead-out terminals
7 = connecting leads of voice coil
8 = spider
9 = front plate

10 = magnetic shorting ring
11 = back plate
12 = magnet
13 = vent
14 = voice coil (speech coil)
15 = voice coil carrier
16 = centre pole
17 = fixing screws

197

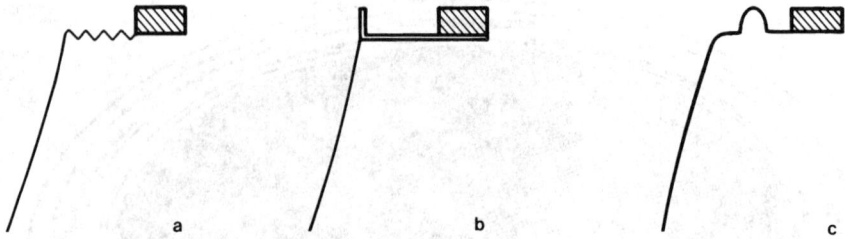

919003-9-17

Figure 4.1.3. Three methods of suspending the cone: (a) pressed-crease
surround for exact centring of the diaphragm even at high loads;
(b) foam rubber surround; (c) roll surround.

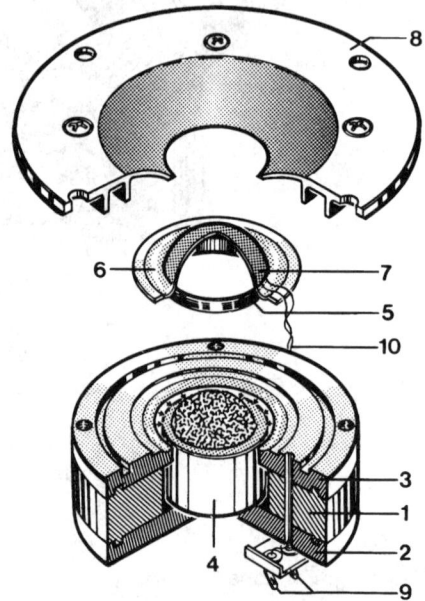

1 = annular magnet
2 = magnet
3 = pole piece
4 = centre pole
5 = voice coil
6 = voice coil carrier
7 = dome
8 = mounting plate
9 = lead-out terminals
10 = connecting lead of voice coil

919003-9-18

Figure 4.1.4. Cross-section of a dome radiator.

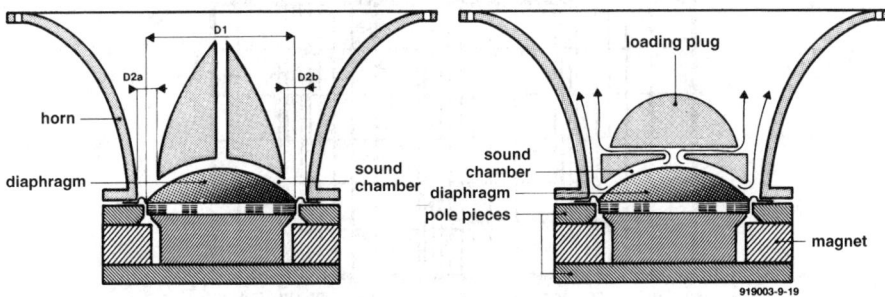

Figure 4.1.5. Principle of a horn-loaded driver with exponential horn.

dome. A cross-section of a dome radiator is shown in Figure 4.1.4.

The dome is made from rigid material (thermoplastic, aluminium) with high internal damping. Its diameter is generally smaller than the wavelength of the sound range to be reproduced so that bunching of frequencies is prevented. Dome radiators are normally used in mid-frequency or high-frequency loudspeakers. Their resonant frequency lies between 200 Hz and 3000 Hz, but this is well below the frequency range for which these drivers are intended.

A full-range driver is essentially a combination of a high-frequency unit and a low-frequency unit mounted on the same chassis. This makes it possible to cover the whole audio range with a single driver. The high-frequency radiator is placed centrally in front of the dust cap of the low-frequency unit. This type of driver is frequently used in monitor loudspeakers.

A horn-loaded driver is a special kind of drive unit. Its operation is clarified in Figure 4.1.5. The diaphragm, driven by a voice coil in a similar way as a standard moving-coil driver forces

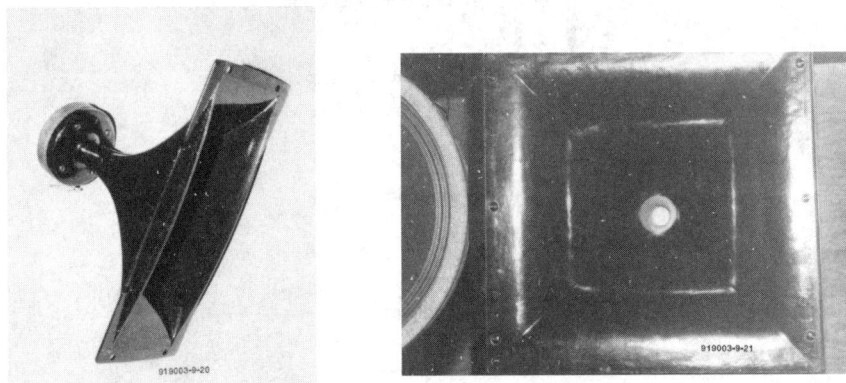

Figure 4.1.6. General view of a horn-loaded driver (left) and view into it (right)

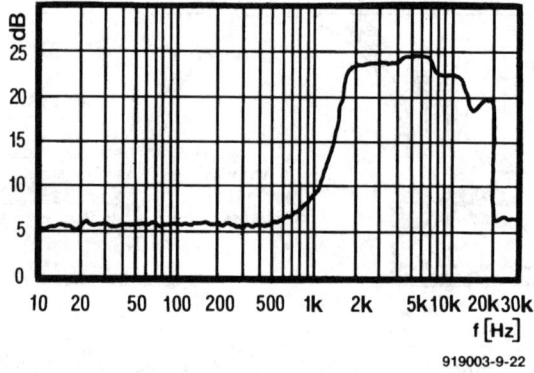

Figure 4.1.7. Frequency response characteristic of a horn driver with exponential horn.

the sound waves through the throat, which increases the velocity of the air flow. This process is called rate transformation. A phase plug is placed in the throat to match it to the surrounding air. Horn-loaded drivers normally have an exponential form—see Figure 4.1.6.

The cross-section of an exponential horn increases exponentially from the throat to the mouth. The horn acts as a high-pass filter. If it is to reproduce bass frequencies, the horn has to be very large. Fir instance, if the lower limit were 50 Hz and the throat diameter 25 cm, the horn would have to be 2.5 metres (9 ft) long. Horn dimensions become acceptable when the lower frequency limit is about 300 Hz. This is why these drivers are used for mid-frequency and high-

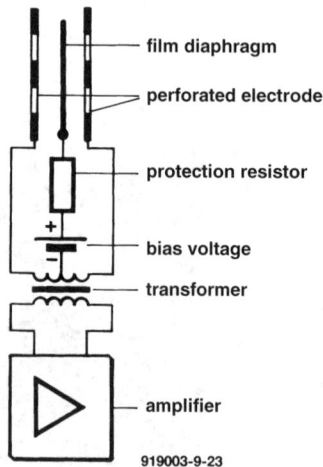

Figure 4.1.8. Principle of the electrostatic drive unit.

200

frequency radiators only. Provided the system is designed well, horn drivers give a very faithful reproduction as shown in Figure 4.1.7.

Horn drivers are really only suited to speech reproduction and as such they find wide application in public-address work in railway stations, airport lounges, and so on. They are not of any interest to musicians.

Piezo tweeters, which are basically inverted crystal microphones, are used for the reproduction of treble frequencies. They consist of a disk of piezo-electric material that is distorted by, and in rhythm with, the alternating voltage applied across it and so radiates sound waves. The resonant frequency of these radiators is fairly high: 4–5 kHz. Below this frequency, no discernible sound is radiated. Nevertheless, piezo tweeters have some advantages: low weight, very high efficiency, and low cost. Moreover, since they have a relatively high internal impedance, they can be connected in parallel with bass frequency drivers without the need of a crossover network.

Electrostatic drivers make use of the force exerted on one another by electric charges to convert electric vibrations into sound. The basic construction of an electrostatic driver is shown in Figure 4.1.8. A thin conducting membrane (1) to which a very high voltage (several kV) is applied, is clamped between two perforated electrodes. The audio signal is applied to these electrodes via a transformer (2). The membrane will be attracted and repelled alternately and starts to vibrate. Since the membrane is very light, these drivers have a very good impulse performance. Owing to their complex construction, they are used as treble radiators only in very expensive hi-fi loudspeakers. Electrostatic drivers are capable of reproducing ultrasonic frequencies up to about 100 kHz.

4.1.1 Specifications

Just as those of other electrical components, the properties of drive unit and loudspeakers are described by a number of parameters. The impedance, Z_{LS} (sometimes erroneously indicated

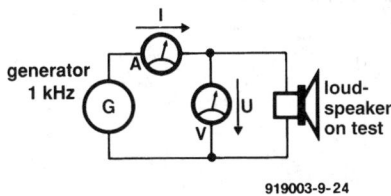

919003-9-24

Figure 4.1.9. Setup for measuring the impedance of the voice coil.

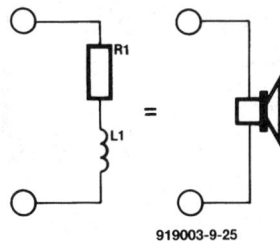

919003-9-25

Figure 4.1.10. Equivalent circuit of an electrodynamic drive unit.

201

by R_{LS}), is the value of the a.c. resistance (see 1.1.4) of the voice coil at a frequency of 1000 Hz. The value specified by the manufacturers has a tolerance of ±10 per cent. Usual values for the impedance are 16 Ω, 8 Ω, and 4 Ω. The actual value of an 8 Ω driver lies between 7.2 Ω and 8.8 Ω. In the vicinity of the resonance frequency this value may be many times greater. The setup for measuring the impedance is shown in Figure 4.1.9. The current I through, and the voltage drop U across, the voice coil are measured. The frequency of the sine wave generator is set to 1000 Hz. According to Eq.9, the impedance is

$$Z_{LS} = U/I$$

In practice, the measurements may be carried out with a good multimeter. The value of the impedance is then approximately

$$Z_{LS} = 1.3R$$

where R is the d.c. resistance of the voice coil measured with the multimeter.

The equivalent circuit of a moving-coil driver in Figure 4.1.10 consists of the series network of a resistor R and and an inductance L. Normally, an amplifier supplies a pure alternating voltage to its output terminals. The reactance of the inductance increases with rising frequency (see 1.1.4). So, the impedance of the loudspeaker consists of an ohmic resistance R and the reactance X_L of the inductance. When a direct voltage is applied to the voice coil, there is no reactance, only a resistance R. This resistance is smaller than the impedance. Of two unequal resistances in parallel, the one with the lower value will have a larger dissipation. This means

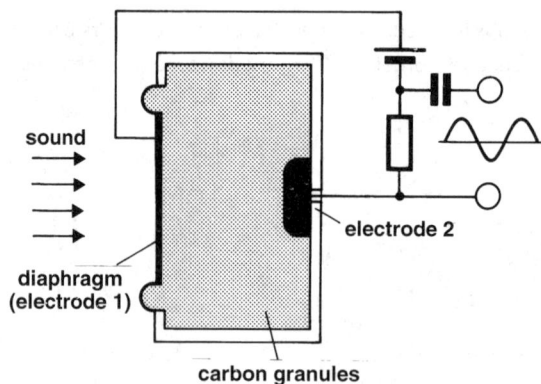

sound

electrode 2

diaphragm
(electrode 1)

carbon granules

919003-3-25

Figure 2.1.10. Principle of a carbon microphone.

202

that in the situation just described, the voice coil would have a larger dissipation than the resist-ance. Ultimately, the voice coil would get warm, then hot, and finally burn out. This shows that a direct voltage always presents a danger to a loudspeaker, and therefore all good-quality power amplifiers have a protection circuit against direct voltages.

The resonant frequency, f, is the eigenfrequency of the vibrating system consisting of voice coil, diaphragm, suspension and surround. The measured value has a tolerance of ± 10 per cent. Actually, a driver has more than one eigenfrequency, but only the lowest, which is the most important, is normally specified.

The frequency characteristic, that is, the sound pressure level as a function of frequency, indicates the frequency range over which the drive unit can be used. The lower and upper lim-its of the range are reached when the sound pressure level has dropped by 10 dB, that is, 32 per cent, referred to the average sound pressure level. Even within the usable range some fre-quencies will sound louder and some softer than others. This implies that the frequency char-acteristic as, for instance, the one in Figure 4.1.11, should be studied carefully before the over-all quality can be judged.

The straighter a frequency characteristic, the more faithful the sound reproduction will be. When studying a characteristic, the scale should be noted, because the vertical scale can be divided into such large segments that any frequency characteristic drawn on it is straight. Some manufacturers provide a hand-drawn frequency characteristic, but most irregularities on this will be unavoidably (or purposely) smoothed out so making the curve look much better than it really is. A properly plotted graph looks like the one in Figure 4.1.11.

The nominal power, P_N, is the maximum power of a sinusoidal signal that can safely be applied continuously to the driver. When this value is exceeded, the drive unit may be damaged, per-haps irrevocably. The test to ascertain the nominal power is carried out with a noise signal. For testing tweeters or loudspeakers containing tweeters, only pink noise should be used, since the amplitude of this drops with rising frequency, just like an actual music signal. The use of other

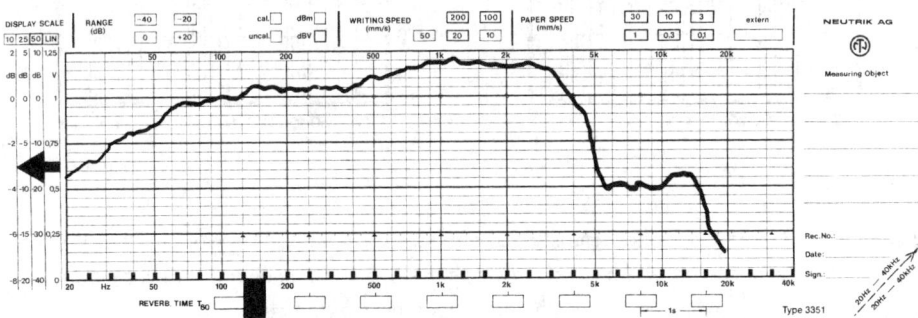

Figure 4.1.11. Frequency characteristic of a bass drive unit.

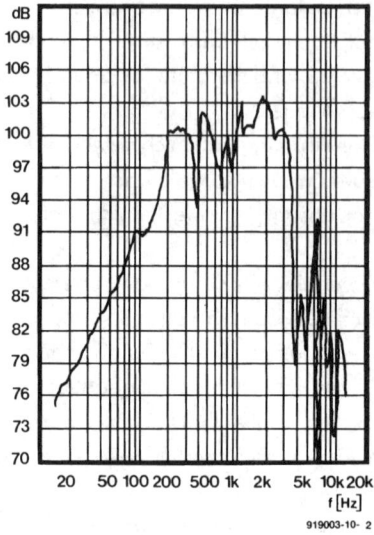

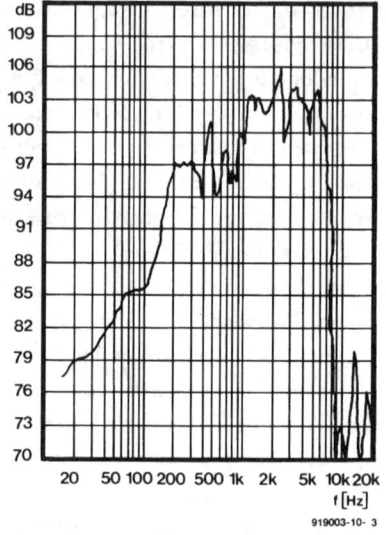

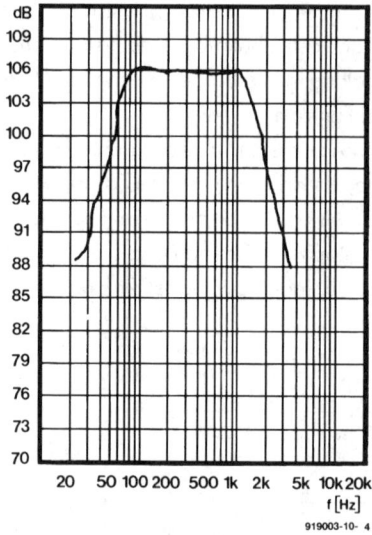

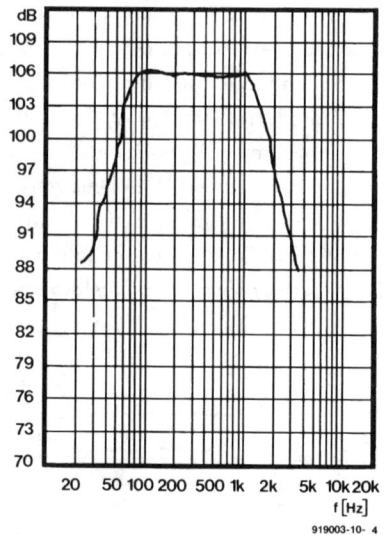

Figure 4.1.12. Frequency characteristics of different drive units. The two curves at the top are actually measured and plotted, whereas the two lower ones are drawn by hand, which is evinced by the smoothness of the curves.

kinds of noise signal might damage the tweeters. Bass drivers can be tested with white noise, however.

In the test, the voltage across, and the current through, the voice coil is measured. The product of these gives the power (Eq.10). The drive unit must be capable of operating properly with a power of P_N watts for an unlimited period of time. When it is mounted in a suitable enclosure, it is capable of handling higher powers. How much higher cannot be ascertained without knowledge of the characteristics of the enclosure. It is, therefore, better to be safe than sorry and not exceed the nominal power rating.

The peak loading rating, P_P, is the peak power that may be applied for short periods without damage to the driver. Note that music power is a worthless piece of information.

The sound pressure level, SPL, is measured in dB at a certain input power and measured at a distance of 1 metre directly in front of the drive unit. Unfortunately, this quantity is sometimes given different names, such as sensitivity, nominal sound pressure or just plain sound pressure. The higher the sound pressure, the higher the sensitivity of the driver. The sensitivity therefore gives some idea of the efficiency of the driver.

Note that the data just described may also be applied to loudspeakers, that is, drivers fitted in suitable enclosures. Example: Figure 4.1.13 shows a moving-coil driver Type E15 from Craft System. According to the manufacturer, this is a 15″ driver suitable for public-address applications and in bass loudspeakers for musicians' amplifiers. The specification of the unit

919003-10-6

Figure 4.1.13. Moving-coil drive unit Type E15 from Craft System.

diameter	15.5 in; 39.5 cm
depth	6.22 in; 15.8 cm
impedance	8 Ω
d.c. resistance	5.2 Ω
continuous music power	300 W
burst peak power	1200 W
sound pressure level 1 W/1 m	103.5 dB
frequency range	35–5200 Hz
resonant frequency	38 Hz
flux density	1.4 T
total flux	2.8 mWb
voice coil material	polyamide
voice coil carrier	aluminium alloy
Q_{ms}	10.3
Q_{es}	0.249
Q_{ts}	0.243
V_{as}	276 litres
weight	8.1 kg

Figure 4.1.14. Technical data of drive unit Type E15 from Craft System.

is given in Figure 4.1.14.

Noteworthy is the large difference between the impedance of the driver and the ohmic resistance of the voice coil (which is measured with an ohmmeter). This indicates that fairly thick wire is used to wind this coil.

The term 'continuous music power' means merely the nominal power rating. The burst (short-term) peak power is given as four times the continuous music power. Since it is not stated how long the burst is, the information is worthless.

The sound pressure level is given for an input power of 1 W at a distance of 1 m and amounts to 103.5 dB. This is relatively high and points to a high efficiency.

The frequency range and resonant frequency speak for themselves. The remaining data refer to the construction of the driver and voice coil and are not of much interest to sound engineers and technicians.

The data in Figure 4.1.14 enable the efficiency of the drive unit to be calculated. First, the radiated acoustic power for an electrical input power of 1 W is calculated by means of Eq.29 (p. 52):

$$L_p = 20\log_{10}[(\rho c W/2\pi r^2)^{1/2}]/p_o$$

which must be rewritten in terms of the acoustic power, P_o

$$P_o = 2\pi l^2 (p_o \times 10^{Lp/20})^2/Z_o =$$

$$= 2\pi \times 1^2 \times (2 \times 10^{-5} \times 10^{103.5/20})^2/408 = 0.138 \text{ W}.$$

The efficiency is then (Eq.11)

$$\eta = P_{out}/P_{in} = 0.138/1 \times 100\% = 13.8\%.$$

This value enables the maximum acoustic power, $P_{o(max)}$, the driver can provide to be calculated. The maximum acoustic power is provided when the applied electrical power is equal to the nominal power rating, 300 W. So,

$$P_{o(max)} = P_N \eta = 300 \times 0.138 = 41.4 \text{ W}.$$

The associated sound pressure level can be calculated with Eq.29 and is 128.3 dB.

In general, Eq. 35 may be used for calculating the efficiency of a loudspeaker, but only if the sound pressure level is measured at a distance of one metre in front of the unit.

$$\eta = 6.16 \times 10^{-10} \times (10^{Lp/10}/P_{in}) \quad [\%] \qquad \qquad \text{[Eq. 35]}$$

where L_p is the sound pressure level in dB and P_{in} the applied electrical power.

Example. A loudspeaker has a voice coil impedance of 8 Ω, the sound pressure level at 1 m in front of it is 90 dB for an electrical input of 1 W, the nominal rating is 240 W, and the frequency range is 50 Hz to 16 kHz. The efficiency is calculated with Eq. 35 as

$$\eta = 6.16 \times 10^{-10} \times (10^{90/10}/1) \approx 0.62 \text{ per cent}.$$

This kind of efficiency is not unusual for a hi-fi loudspeaker covering the whole audio range. The lower the efficiency, the mellower the sound at a given amplifier power output. To obtain a high sound pressure level, the amplifier has to provide quite a lot of power and, clearly, the loudspeaker must be capable of handling this.

4.1.2 Polar diagram

The polar diagram of a loudspeaker was discussed briefly in Section 1.3. If no measures

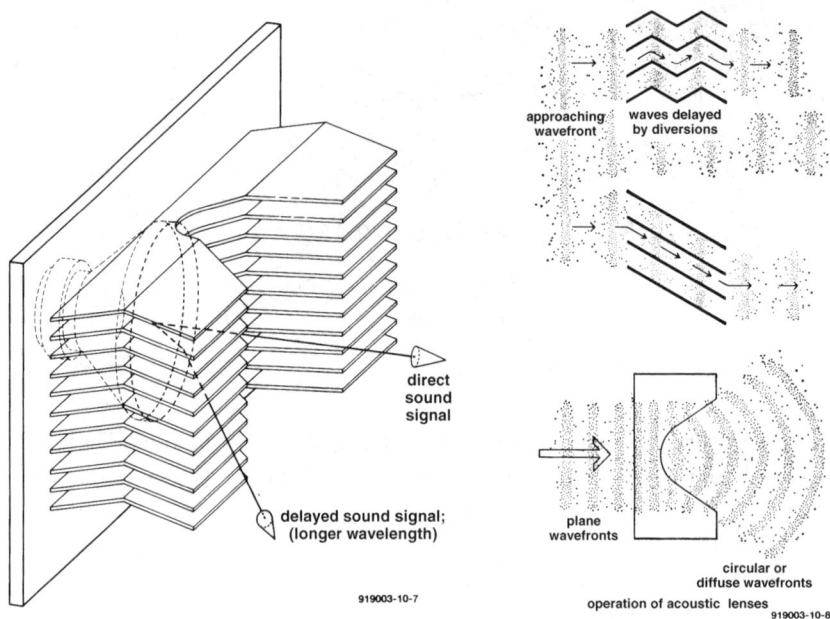

Figure 4.1.15. Acoustic lenses disperse sound waves in different directions. Their action is similar to that of concave optical lenses.

are taken, high frequencies become bunched. The lower the frequency, the more the polar diagram becomes omnidirectional. The reason for this is that at high frequencies the wavelength of the sound is comparable to, or smaller than, the diameter of the diaphragm of the drive unit. This is why manufacturers always state the frequency response only in a straight line in front of the driver or loudspeaker. However, the directivity of a loudspeaker is not determined by its frequency response alone. It matters also whether the loudspeaker uses a single driver or several. If the latter, the angle of radiation is larger owing to interference phenomena. To reduce the bunching at high frequencies, acoustic lenses are used that are placed directly in front of the loudspeaker(s). These disperse the sound waves in a similar way as concave optical lenses disperse light. The operation of acoustic lenses depends on the fact that the propagation velocity of sound waves is lower in the lens than in free space. The principle of such a lens is illustrated in Figure 4.1.15.

When the loudspeakers are being positioned on the stage, differences in their directivity must be taken into account. As far as an electric bass guitar is concerned, it does not matter where the loudspeaker is placed since at low frequencies the sound source cannot be determined in any case. However, for other instruments it does make a difference

208

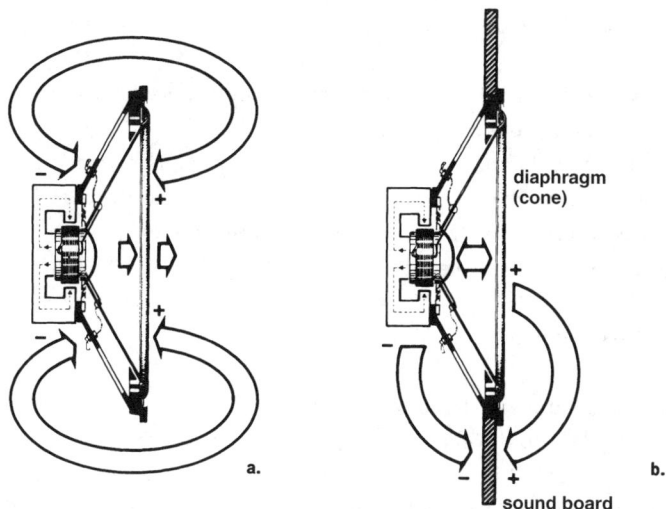

Figure 4.1.16. Acoustic short-circuit (a) and how to obviate it by means of a baffle (b).

where the loudspeakers are placed. To distribute the sound evenly and produce a spatial sound field, the loudspeakers of two equivalent instruments, such as rhythm guitars, must be not too close together if at all possible.

A drive unit cannot radiate low frequencies owing to a phenomenon called acoustic short-circuit. When a low-frequency signal is applied to the unit, the diaphragm moves forward owing to the current through the voice coil. The air in front of the diaphragm is then condensed, but that behind it, rarefied. Since at low frequencies sound waves are propagated spherically, those radiated at the front may travel to the back and vice versa (see Figure 4.1.16). The differences in air pressure at the front and rear cancel each other so that there is no net sound pressure in spite of the vigorously moving diaphragm.

The effect of acoustic short-circuiting reduces with increasing frequency since high frequencies are bunched and therefore do not travel to the back of the diaphragm. The only effective way of obviating an acoustic short-circuit is mounting the driver on a baffle or in an appropriate enclosure. In general, the larger the baffle, the lower the frequencies that are reproduced. With an infinitely large baffle, the low-frequency limit is determined by the eigenfrequency of the driver. The maximum possible acoustic energy is radiated when the diameter of the baffle is larger than the longest sound wave to be reproduced. So, the lowest frequency of a bass guitar (41.2 Hz) is radiated when the diameter of a circular baffle is

$df = c/f = 343/41.2 \approx 8.3$ m

If a reduction of 2 dB (or 20%) in sound pressure level is acceptable, the diameter can be reduced by two thirds, but even then will still be 2.7 m.

A driver should not be mounted at exactly the centre of, or symmetrically on, the baffle, since this gives rise to standing waves that may cause one or more frequencies of the sound to be suppressed partly or even entirely. The frequency response is considerably better when the driver is mounted off-centre on the baffle.

4.1.3 Multi-way loudspeakers

Many loudspeakers use two, three or even more drive units to obtain a straight frequency response. Each of these drivers radiates part of the audio range only (bass, treble, mid-frequency, and so on). It is, of course, important to ensure that the total impedance of such a combination does not fall below the output impedance of the amplifier and that the rating of the individual drivers is not exceeded. It is also essential that the drivers are connected acoustically in phase to prevent acoustic short-circuits. Acoustically in phase means that the diaphragms off all drivers move synchronously forward or backward in line with the applied voltage. If the drivers are used without a crossover network (or filter), this can be achieved only by linking the terminals of the drivers in the correct way as shown in Figure 4.1.17.

If there are markings on the drive unit terminals, the positive one is normally indicated by a red dot. If there are no markings, briefly connect a 1.5 V battery across the speech coil and note into which direction the diaphragm moves. If it moves forward, the +ve ter-

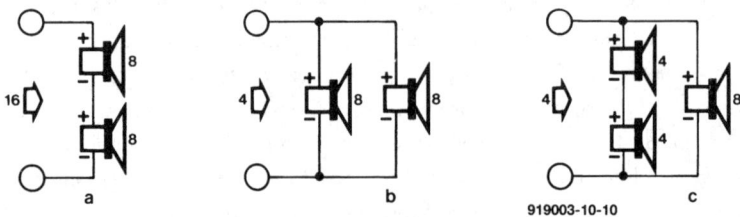

919003-10-10

Figure 4.1.17. Combining a number of loudspeakers without a crossover network, for instance, several woofers linked to a single instrument amplifier. In (a) two 8-Ω drivers are connected in series to give a total impedance of 16 Ω. In (b) the two are connected in parallel so that their total impedance is 4 Ω. (c) shows a typical series-parallel combination of three drivers. The two series-linked 4 Ω units have a combined impedance of 8 Ω, which, shunted by the single 8 Ω unit, gives a total impedance of 4 Ω.

210

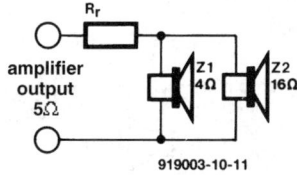

Figure 4.1.18. How to connect two drivers with dissimilar impedance to an amplifier.

minal of the battery is linked to the +ve terminal of the driver.

Drivers may be combined in three ways: series, parallel or series-parallel. The same rules apply as with resistances (1.1.9). Superfluous power from an amplifier may be dissipated in a power resistor. For instance, a 4 Ω driver rated at 20 watts in combination with a 16 Ω driver also rated at 20 W, must be linked to an amplifier rated at 30 W with an output impedance, Z_a, of 5 Ω. A possible way of doing this is shown in Figure 4.1.18. The series resistor compensates for the mismatch between the driver combination and the amplifier, and also dissipates the unwanted power of 10 W. The total impedance, Z_i, of the driver combination is

$$Z_i = Z_1 Z_2/(Z_1 + Z_2) = 4 \times 16/20 = 3.2 \ \Omega.$$

The value of the series resistor must be the difference between this impedance and that of the amplifier output:

$$R_r = Z_a - Z_i = 5 - 3.2 = 1.8 \ \Omega.$$

The amplifier delivers a current, I, which, according to Eq. 10, is

$$I = \sqrt{(P_a/Z_a)} = 30/5 = 2.45 \ A.$$

Since this current flows through resistor R_r, the power dissipated in this resistor is

$$P = I^2 R = 2.45^2 \times 3.2 = 19.2 \ W.$$

This means that the series resistor should be a vitreous type rated at 20 W.

The voltage, U_{LS}, across the drive units is

$$U_{LS} = I Z_i = 2.45 \times 3.2 = 7.84 \ V.$$

211

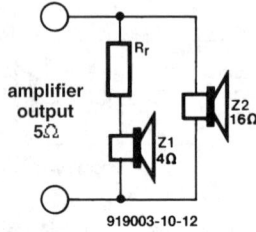

919003-10-12

Figure 4.1.19. Another way of connecting the speakers in Figure 4.1.18.

The power across the 4 Ω driver is

$$P_{4\Omega} = U_{LS}^2/Z_1 = 7.84^2/4 = 15.37 \text{ W},$$

and that across the 16 Ω driver is

$$P_{16\Omega} = 7.84^2/16 = 3.84 \text{ W}.$$

This shows that neither of the drivers is overloaded.

Another way of connecting the drivers is shown in Figure 4.1.19. The 16 Ω drive unit is in parallel with the series network of the 4 Ω driver and series resistor R_1. In this case,

$$Z_i = Z_2(Z_1 + R_1)/(Z_1 + R_1 + Z_2),$$

from which

$$R_1 = [16Z_i/(16 - Z_i)] - 4 = 3.3 \text{ } \Omega.$$

The potential, U_{LS}, across the combination is

$$U_{LS} = \sqrt{(P_a/Z_a)} = \sqrt{(30/5)} = 12.25 \text{ V}.$$

The potential across R_1 is

$$U_R = U_{LS}R_1/(R_1 + 4) = 12.25 \times 3/(3.3 + 4) = 5.54 \text{ V},$$

and the consequent power dissipated in it is

212

$P_R = U_R^2/R_1 = 5.54^2/3.3 = 9.29$ W.

The power dissipated in Z_2 is

$P_{z2} = U_{z2}^2/Z_2 = 12.25^2/16 = 9.38$ W.

The potential across Z_1 is $12.25 - 5.54 = 6.71$ V and the power dissipated in it is $6.71^2/4 = 11.26$ W.

In this alternative circuit also, neither of the drivers is overloaded, but the power distribution across them is better.

When bass, treble, and mid-range drivers are combined in an enclosure, a crossover filter has to be used to split the overall frequency range into appropriate bands. The simplest such filter is a potential divider as shown in Figure 4.1.20, in which the relevant driver acts as part of the divider. The circuit is a high-pass filter whose response is shown in Figure 4.1.21. The frequency f_c at which the signal level is down by 3 dB with respect to

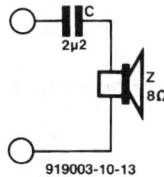

Figure 4.1.20. High-pass filter formed by a capacitor and the inductance and resistance of the voice coil of the drive unit.

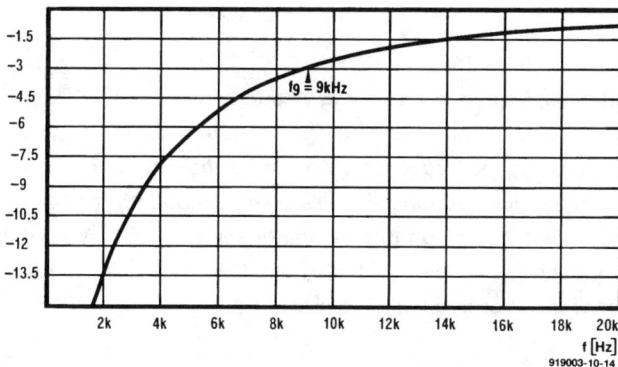

Figure 4.1.21. Transfer characteristic of the high-pass filter in Figure 4.1.20.

213

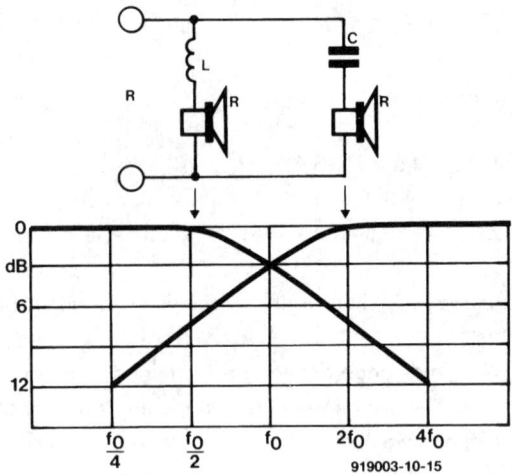

Figure 4.1.22. A bass driver and a treble driver linked by a crossover network formed by a capacitor and inductor and below it the transfer characteristic of the combination.

the maximum level is called the cut-off frequency. It is calculated with

$$f_c = 1/2\pi RC \qquad\qquad \text{[Eq. 36]}$$

A low-pass filter is used to attenuate bass frequencies above a certain point, that is, the cut-off frequency. It normally consists of inductors and resistors, and its cut-off frequency is calculated with

$$f_c = 1/(2\pi L/R) \qquad\qquad \text{[Eq. 37]}$$

A crossover network as shown, together with its frequency response, in Figure 4.1.22 consists of a low-pass filter and a high-pass filter. The attenuation is 6 dB/octave (an octave is a frequency interval in which the frequency is doubled or halved). In Figure 4.1.22, the filters have the same crossover frequency (as the cut-off frequency in a crossover filter is called). It is important that the roll-off (that is, the attenuation) is a straight line. In practice, the crossover frequencies are determined by the drive units. The inductance, L (in millihenries – mH), of the inductor in the low-pass filter is

$$L = 160R/f \qquad\qquad \text{[Eq. 38]}$$

214

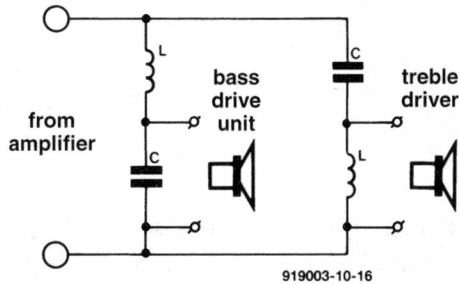

919003-10-16

Figure 4.1.23. Circuit of a crossover filter similar to the one in Figure 4.1.22, but having a (steeper) roll-off of 12 dB/octave.

where R is the impedance of the relevant voice coil and f is the cut-off or crossover frequency. The capacitance, C (in microfarads – μF), of the capacitor in the high-pass filter is

$$C = 16 \times 10^4 / fR \qquad\qquad \text{[Eq. 39]}$$

The simple crossover filter just considered has an attenuation or roll-off of 6 dB/octave, and is called a first-order filter. There are 2nd, 3rd, and 4th order crossover filters with roll-off characteristics of 12 db/octave, 18 dB/octave, and 24 dB/octave respectively. The greater the roll-off, the steeper the skirts of the frequency response characteristics and the sharper the crossover. High-order filters are needed when a loudspeaker has a linear response over only a limited frequency band. The circuit of a 2nd order crossover network is shown in Figure 4.1.23. The values of the relevant component are calculated with

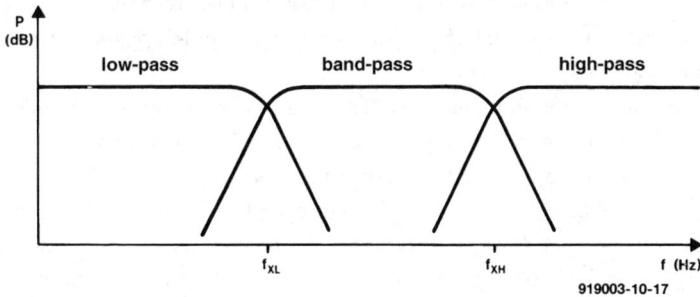

919003-10-17

Figure 4.1.24. In good loudspeakers, the overall frequency range is divided into three or more bands, each reproduced by a discrete driver.

215

$$L = 225R/f \ (\text{mH}) \hspace{4cm} [\text{Eq. 40}]$$

and

$$C = 112 \times 10^3/fR \ (\mu F) \hspace{4cm} [\text{Eq. 41}]$$

Inductors are always air-cored types to prevent distortion owing to hysteresis phenomena in the (iron) core. The capacitors are invariably low-loss types (MKP=polypropylene; MKC=polycarbonate; MKT= polyester).

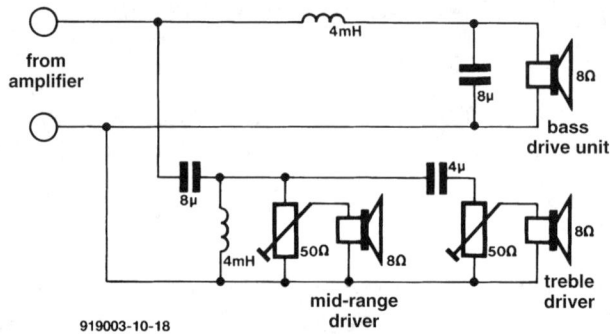

Figure 4.1.25. Circuit of a three-way crossover filter with two cross-over frequencies.

When only one frequency is stated, the filter is a two-way type. In hi-fi and studio loudspeakers, the audio frequency range is divided not into two, but into three or more bands as shown in Figure 4.1.24. The crossover filter must then be designed for three frequency bands with two crossover frequencies. The circuit of such a filter is shown in Figure 4.1.25. Note that the sound pressure level of the mid-range and treble drivers can be adjusted with preset potentiometers.

It should be borne in mind that the phase response of a loudspeaker is modified by the crossover filter. It is, therefore, necessary when a crossover is used to connect the bass and mid-range units so that they, if used without a filter, would be in anti-phase. In other words, the polarity of the mid-range driver is the opposite of that of the bass driver. The polarity of the tweeter (treble driver) is immaterial. Proprietary filters are normally marked with the polarity (+ or –) of each of the drivers to minimize the risk of mistakes.

The quality of proprietary crossover filters varies widely. The enamelled copper wire of the inductors must be heavy-duty to minimize losses. A good-quality passive filter reduces the efficiency of the loudspeaker only slightly: by about 1–2 dB. As mentioned earlier, the

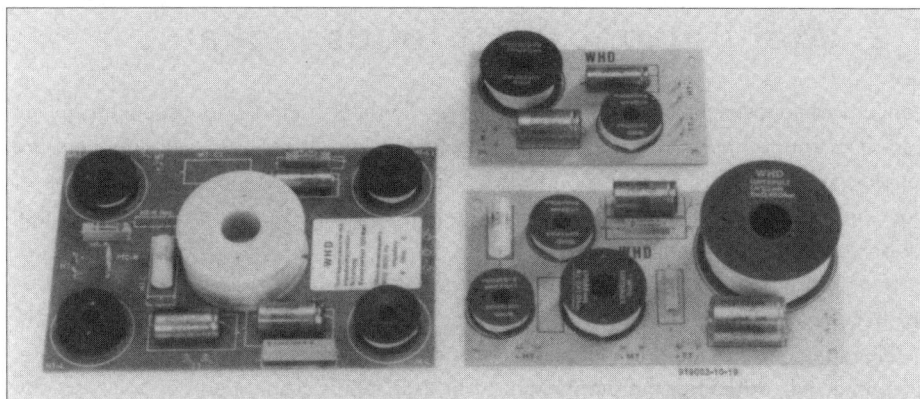

*Figure 4.1.26. Commercially available crossover filters
for use in DIY loudspeaker enclosures.*

inductors must be air-cored as iron or ferrite cores degrade the sound quality appreciably.

Each high-pass of low-pass filter is designed for a specific loudspeaker impedance and cannot be used with loudspeakers with a different impedance. If, say, a 16 Ω tweeter would be linked to a 4 Ω high-pass filter, the cut-off frequency would shift and the frequency response would become uneven.

The filter must be capable of dissipating the entire power applied to the loudspeakers. When the loudspeaker is used for normal music reproduction, the power sharing by the bass, mid-range and treble drivers is not uniform. This is so because in such reproduction the amplitude drops as the frequency rises. This means that the mid-range and treble drivers get only a small part of the total power available. How large this part is depends on the crossover frequencies. The higher these are, the lower the power applied to the mid-range and treble units. It is clear that the rating of a loudspeaker is that of the bass driver. It also explains why howl caused by positive feedback can be fatal for a treble driver. If, for instance, the frequency of the howl lies above the relevant crossover frequency, the treble unit suddenly has to handle the full output power, which it is not designed to do. The inevitable consequence is that the unit gives up the ghost. This is why some manufacturers fit fuses in the supply leads to the mid-range and treble units. But since these are very slow types to prevent them blowing with normal high-volume music, they often blow only after the treble unit has been damaged beyond repair. However, they do protect against the risk of fire.

4.2 Monitor and hi-fi loudspeakers

Monitor loudspeakers as used in the recording studio differ from hi-fi loudspeakers only in respect of shape and rating: most other parameters are the same. Such speakers are housed in closed boxes, so that sound waves leaving the enclosure at the front can never reach the back of the diaphragm. Acoustic short-circuits can, therefore, not occur so that the box behaves as an infinite baffle.

This does not mean that other problems cannot arise. One of these is that air in the box becomes condensed and so damps the movement of the diaphragm. Consequently, the eigenfrequency of the driver rises in inverse proportion to the volume of the enclosure. To counter this damping, the diaphragm of hi-fi drivers is suspended flexibly. Also, the mass of the diaphragm/voice coil system (the motor) is increased. Another problem may arise from the tendency of the box to self resonance. All this means that the design of a good enclosure is difficult, which is reflected in the price and cost, which vary enormously from one loudspeaker to another.

Apart from closed boxes, there are vented boxes, normally called bass reflex boxes. In these, the sound radiated backwards by the driver is turned forward by a very accurately calculated bass reflex vent, which greatly improves the bass reproduction. The enclosure is calculated on the basis of the eigenfrequency of the bass driver.

To a sound technician, normally only the electrical parameters of the loudspeakers are of interest. Monitor and hi-fi loudspeakers must be capable of reproducing the entire audio range with as little distortion as feasible, since their reproduction is the criterion by

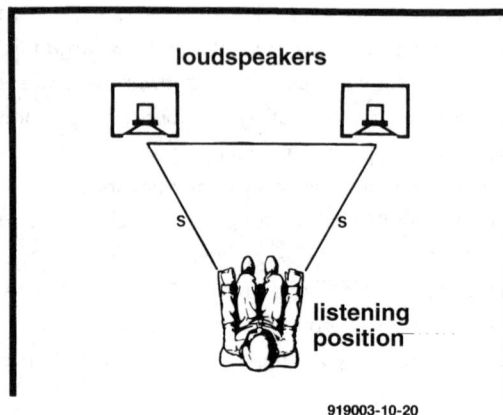

919003-10-20

Figure 4.2.1. The best stereo sound is obtained when the two loudspeakers and the listener form an equilateral triangle.

which the musical end result is judged. In spite of all this, it remains a fact that the loudspeaker remains the weakest link in the sound reproduction chain.

Monitor loudspeakers must be positioned carefully, since if they are not, it will be difficult to judge the quality of the reproduced sound. They must be at ear-height to ensure that the proportion of the direct sound that reaches the ear is as large as possible. Also, they must be placed in neutral positions; for instance, when they are put in a corner, the bass frequencies are emphasized unnaturally. The best position is on a free-standing plinth. If at all possible, the loudspeakers and the listener(s) should form an equilateral triangle – see Figure 4.2.1. In this way, optimum stereo performance is ensured at the listening position, which is, for instance, behind the mixer.

4.3 Voice loudspeakers

Voice loudspeakers were already discussed in Chapter 3. Nowadays, these loudspeakers are intended not just for the vocal parts of the performance, but also for a number of acoustic instruments. Consequently, the requirements of these units are virtually identical to those of monitor loudspeakers: for instance, they must be capable of reproducing the entire audio frequency range of 40–16000 Hz. Since they are transported regularly, they must be robust, proof against overload, and have a high continuous rating. The drive units invariably have a rigidly suspended diaphragm. A voice loudspeaker with a bass driver and two piezo tweeters is shown in Figure 4.3.1.

Figure 4.3.1. Passive voice loudspeaker consisting of a bass driver and two treble drivers. Because of the use of piezo tweeters, a crossover filter is not needed.

219

Apart from passive loudspeakers, there are also active types, in which the output amplifier and crossover filters have been incorporated in the same enclosure as the drive units. High-quality active loudspeakers have a separate output stage for each driver. The frequency range is then divided into two or three frequency bands by a crossover filter preceding the output stages.

4.4 Instrument loudspeakers

Musical instruments are normally reproduced by one or more equivalent loudspeakers, usually bass or wide-band types. Since these speakers have to be rugged, the diaphragm of the drivers is normally rigidly suspended. Owing to the large sound pressure levels, the panels of the enclosure vibrate in unison with the reproduced sound, which clearly affects its colouring.

Instrument combinations are frequently provided with two loudspeakers, but discrete instrument loudspeakers always use two or more drivers. The guitar loudspeaker from Marshall has four identical drivers Type G12 from Celestion and reproduces the mid-range and treble frequencies completely differently than a loudspeaker using only one of these drivers.

Apart from the directivity, the sound of the discrete drivers also influences the final sound to a large extent. Judging this in the case of instrument loudspeakers and combined systems is quite subjective. Studying and comparing frequency response characteristics is of little use. Before a loudspeaker is purchased, it should be tested with one's own amplifier and instrument(s). This is especially advisable where guitar loudspeakers are concerned.

Many guitar loudspeakers are open at the back, which improves the efficiency of the unit appreciably. It makes it possible for very high sound pressure levels to be produced with a relatively modest amplifier output. The frequency response characteristic of these loudspeakers is, of course, not straight, but that does not say much about the quality of the sound. Normally, bass drivers with a relatively high eigenfrequency (60–100 Hz) and covering an audio band up to about 5 kHz are used for the reproduction of the guitar signal. Since most guitar amplifiers emphasize the mid-range and treble frequencies, the sound is clear enough for most guitarists. It is rare for wideband drivers to be used in guitar loudspeakers. The use of special treble units would degrade rather than enhance the sound, since this would become very sharp. The frequency response would also be unusual and not at all to the liking of many listeners.

Some loudspeakers for electric bass guitars have a bass reflex vent to improve the efficiency of the loudspeaker. In accordance with the lowest tone produced by an electric bass guitar, most loudspeakers used with this instrument have an eigenfrequency of 30–40 Hz.

The upper frequency limit is usually 3000–5000 Hz. Since a bass guitar does not produce tones above 4000 Hz, the use of a wideband loudspeaker would not have any effect on the final sound.

In the case of keyboard instruments, the frequency response of the loudspeaker is more important, because these instruments nowadays produce a ready sound, which means that the loudspeaker(s) must meet the same requirements as hi-fi loudspeakers. In fact, during large concerts, the keyboard loudspeaker often functions as the monitor speaker.

Leslie loudspeakers are in a different category: they are normally and preferably used with electric organs and keyboards. They use rotating drivers, which results in a distinct, attractive sound.

When buying a loudspeaker, always check that its rating is equal to, or preferably larger than, the rating of the output amplifier with which it is to be used. Since instrument amplifiers are often overloaded to permissible limits, it is advisable to choose a loudspeaker with a much higher rating than the amplifier (30 per cent or more).

4.5 Public-address loudspeakers

Loudspeakers with high ratings are normally exponential types, whose efficiency is higher and whose distortion is lower than those of standard loudspeakers. The consequent gain in sound pressure level can be as much as 9 dB, equivalent to an eight-fold rise in amplifier output. So, in large PA systems, the use of exponential loudspeakers leads to a cost reduction: fewer loudspeakers are needed and the amplifier output can be smaller than needed with conventional loudspeakers.

The design of exponential loudspeakers is complicated. There are three main types—see Figure 4.5.1:

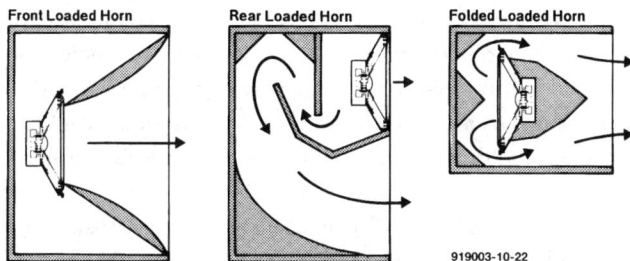

Front Loaded Horn   Rear Loaded Horn   Folded Loaded Horn

919003-10-22

Figure 4.5.1. Three types of exponential loudspeaker.

Figure 4.5.2. This mid-range/treble loudspeaker, used by a well-known rock band, consists of a front-loaded horn, two high-frequency horns, and four tweeters.

- Front-loaded horn, which is the original exponential loudspeaker. A horn, whose diameter increases exponentially in a forward direction, is placed in front of the drive unit. Since the dimensions of a horn for low frequencies would be unmanageably large, this type of unit is used only for mid-range applications (see Figure 4.5.2).
- Folded horn, in which, as the name implies, the horn is folded over to six to ten times. The design is very complicated. Folded horns are used for the reproduction of bass frequencies.
- Rear-loaded horn, which is a combination of a standard loudspeaker and an exponential horn. The loudspeaker radiates mid-range frequencies forward in the traditional manner. At the rear, bass frequencies are reproduced by the folded horn.

4.6 Home construction of loudspeaker enclosures

Home construction of loudspeaker enclosures for hi-fi and studio recording purposes is not advisable, particularly in the case of bass reflex enclosures, and certainly not if success is to be guaranteed beforehand. Construction is highly complicated and requires extensive knowledge, experience, competence and relevant measuring and test facilities. Instrument loudspeakers, however, can be built at home without too much trouble; all that is needed is competence with hammer and saw. It is worthwhile doing, because quite a lot of money can be saved: loudspeakers offered in the retail trade are generally highly overpriced.

There are no guidelines for the home constructor, although there are several good books and magazines available dealing with the construction of loudspeaker cabinets. Suffice it to say that the enclosures should be sturdy and capable of withstanding some rough treatment. It is strongly recommended to choose the loudspeaker rating a good 30 per cent or more above the rating of the amplifier used. Enclosures are normally constructed from 18 mm, 22 mm, or even 25 mm medium-density fibreboard (MDF). The panels are fixed together with good-quality wood glue and screws. The drive units should be fitted to the relevant pane (normally the front) with nuts, bolts and washers. Transporting the speakers may be facilitated by the fitting of suitable handles and castors (or small wheels).

Finally, instead of buying the latest available drivers, try to purchase drivers whose production has been discontinued: these are normally much cheaper (and still good units) than the latest models.

5. Peripheral & effects units

Apart from instruments, amplifiers and loudspeakers, a band needs a galaxy of ancillary equipment, such as effects units, footswitches, pedals and a large number of cables and leads.

5.1 Effects units

The music and electronics retail trade offers a bewildering variety of ancillary and effects units that can be inserted into the signal path between instrument and amplifier or linked to the effects input and output of the mixer or amplifier. These units make possible the production of innumerable sound effects. Normally, they are operated via a footswitch and powered by an external mains adaptor. They rarely have their own power supply.

It is not possible in a book of this nature to discuss all the effects units currently available, added to which new models appear on the market regularly. Fortunately, the properties and general design specifications of these units do not vary all that much. Moreover, the use of effects units must remain a question of taste (or lack thereof).

5.1.1 General

Peripheral and effects units are n o different from other electronic circuits inserted into the pathway of the signal: they reduce the dynamic range and cause the noise level to rise. Also, the reliability of the overall system suffers, because each unit adds components and parts that may fail. In addition, each unit adds at least two plug-and-socket connections which can cause problems (transfer resistance, bad contacts, and others). It is clearly advisable to restrict the effects equipment to those units that are really needed. It is also recommended to use units that have a low noise figure.

The electronic circuits needed to produce an effect are not very complicated, but the components used determine the quality and reliability (and the price: low-noise semiconductors, gold-plated contacts, dustproof and crackle-proof potentiometers, and so on). When buying an effects unit, a listening test, preferably with one's own amplifier and instrument, is an absolute must.

When deciding on whether to use effects units or not, the overall sound must be borne in mind at all times. Too many or overemphasized effects become predominant and contaminate the original sound. So, effects units should be used sparingly and carefully. As in so many other matters: the fewer, the better.

An effects unit may be inserted between instrument and amplifier or mixer or to the relevant input and output of the amplifier (between preamplifier and output stage). Units of the first category are invariably operated by a footswitch or footpedal and are usually battery powered. Mains operation is not advisable owing to the added cables across the stage and the increased risk of mains hum. Note, by the way, that although most of these units are provided with a footswitch, they are only turned off when the connecting cable is pulled from the socket. So, to save the battery, always pull out the connecting cable during intervals.

Units of the second category are normally housed in a 19 in enclosure which fits in a rele-

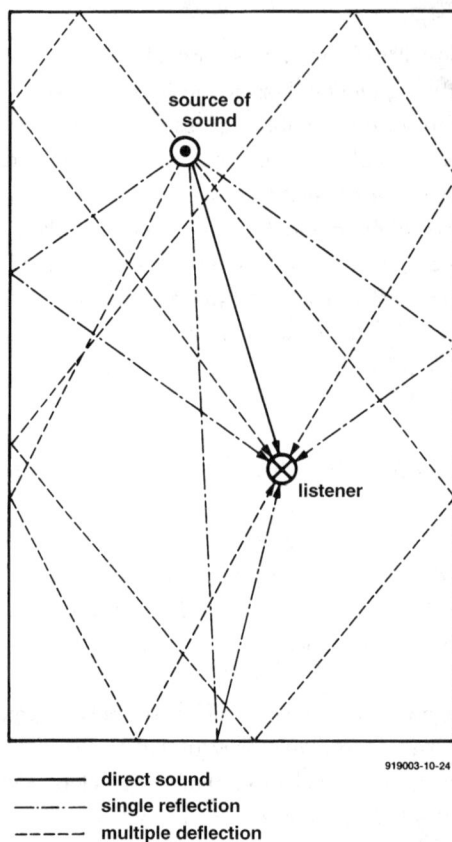

919003-10-24

——————— direct sound
—·—··— single reflection
– – – – – multiple deflection

Figure 5.1.1. In an enclosed area, sound reaches our ears not only directly, that is, via the shortest possible route. Multiple reflections from walls, ceilings, and furniture reach our ears with some delay. The reflected sound waves cause the reverberation.

226

vant rack. These units are normally mains powered and usually have more and better facilities than their battery-operated counterparts. However, their electro-acoustical properties are not necessarily better.

When an effects unit is inserted between the instrument and amplifier, only the original signal of the instrument can be influenced. If, however, it is linked to the relevant input and output of the amplifier, the total signal processed by the amplifier may be influenced. Many vocal and musician's amplifiers incorporate an effects unit as standard. It is also possible to build an effects unit into an existing equipment by means of a construction kit, of which there are a number on the market.

5.1.2 Spatial effects

To make an audio signal give a spatial impression, use is made of reverb(eration) and delay lines. These are undoubtedly the most important effects that are added to the dry sound, which then becomes fuller and gives a clear spatial impression. Reverb and delay are quite different effects which should not be confused. Reverberation is caused by multiple reflections of the sound against walls, ceilings and furniture. These reflections are delayed by 10–100 ms. The total time lapse before all reflections have sounded is called the reverberation time. Natural reverberation times vary from 200 ms (small halls) to 5 s (churches and cathedrals).

Reverberation may be produced artificially in a number of ways. The reflected sound waves reach our ears at irregular intervals of time as shown in Figure 5.1.1. This makes producing a

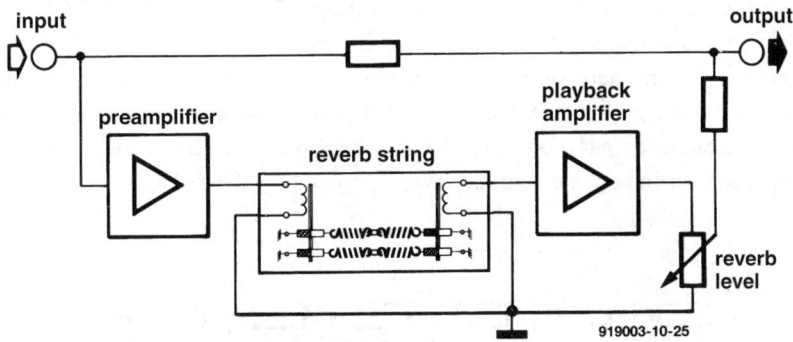

Figure 5.1.2. Simplified block diagram of a spring line reverb unit. The signal is first amplified and then applied to a converter which sets the spring into vibration. At the other end of the spring, the mechanical vibrations are transformed by a second converter into electrical vibrations, which are magnified by a playback amplifier. The original input signal is also applied to the output via a resistor. The ratio of signal plus reverb to signal without reverb is set with the reverb level potentiometer.

natural sounding reverberation very difficult, especially in the case of digital equipment. Note that the quality of reverberation so produced can be judged by ear only.

There are analogue reverb units (using spring lines or bucket brigade memories) and digital ones. In some studios, reverberation is produced by a number of microphones in a diffuse sound field in a well-reflecting recording area, but such facilities are rare. The type of reverb unit to be obtained depends, of course, on the application.

In a spring line reverb unit, signals are introduced at one end of the spring and picked up at the other. The time delay caused by their length along the spring creates the effect: the longer the spring, the deeper the reverb. Feedback is normally also provided to enhance the effect. The frequency range is limited to the midrange (a couple of hundred Hz to a few kHz). It is advisable to use this type of reverb unit only with a musician's amplifier. In many combos, a spring line is incorporated. It is obvious that the spring(s) must be screened from extraneous sounds to prevent it being set into spurious vibration. A simplified block diagram of a spring line reverb unit is shown in Figure 5.1.2

Another means of producing analogue reverb is the bucket brigade memory shown in Figure 5.1.3. A reverb unit of this design is immune to mechanical shocks and sound waves. The reverberation is produced entirely electronically. Unfortunately, this type of unit can be used only with midrange frequencies, but they are appreciably less expensive than spring line types. Also, for use with musician's amplifiers, they are an inexpensive alternative to digital reverb units, which are now standard in recording studios.

Digital reverb units are much more complicated than analogue types. The analogue audio input signal is first sampled and then digitized. The sampling rate must be at least twice as high as the highest audio frequency to be reproduced. So, if audio frequencies of up to 15 kHz have to be processed, the sampling rate must be at least 30 kHz.

The data stream so produced is delayed by a microcomputer and then reconverted into an analogue signal (see Figure 5.1.4). The parameters of the desired reverb are input into the computer and processed by its program (software). This type of reverb unit has a much larger frequency range than analogue units and can produce a galaxy of reverberation effects. Apart

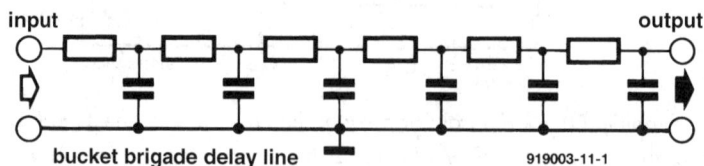

Figure 5.1.3. Principle of the delay produced by a bucket brigade memory. Owing to the sequence of resistors and capacitors, the input signal arrives at the output after a certain time delay.

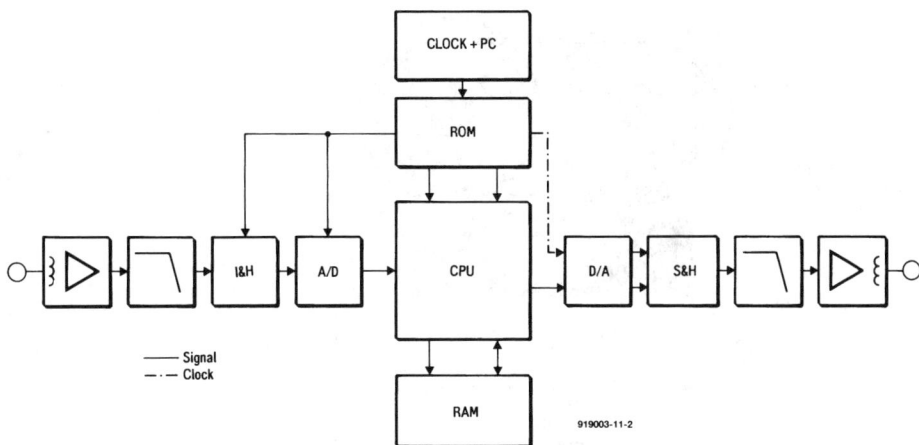

Figure 5.1.4. Block diagram of a basic digital echo or reverb unit. The input signal is
passed through a preamplifier and a low-pass filter before it is sampled and
quantized. A microcomputer (ROM, RAM and CPU) provides the delay or effect.
After this, the digital data are reconverted into an analogue signal.
All operations are carried out sequentially in synchrony with a clock signal.

from the frequency range, the number of simulated reflections per second is a criterion of their quality. Unfortunately, this property is seldom specified by the manufacturer For a natural sounding reverberation, a great number of reflections per second with irregular intervals is a must.

In contrast to a reverb unit, an echo or delay unit only repeats the signal, which gives information about the distance to, and the nature of, the reflecting walls. Echoes cannot be produced naturally, but can readily be simulated by technical means. The principle is always the same: the signal is delayed appreciably and then combined with the original signal. Normally, the delay time, depth of the effect, and the number of repeats can be varied. Today, there are a number of ways of producing echoes. The simplest of these is by the use of a bucket brigade memory, which, as before, is unfortunately only usable with midrange frequencies. Nevertheless, this type of unit is perfectly suitable for use with an electric guitar and offer an inexpensive means of producing an analogue delay.

Tape echo equipment as shown in Figure 5.1.5 is much more complicated and was much more often used in the past than nowadays. In it, an endless tape runs along a number of tape heads. The first of these is an erase head, followed by a recording head, and one or more playback heads. The time delay depends on the tape speed and the spacing between the recording head and playback head(s). The delay time, Δt, is given by

229

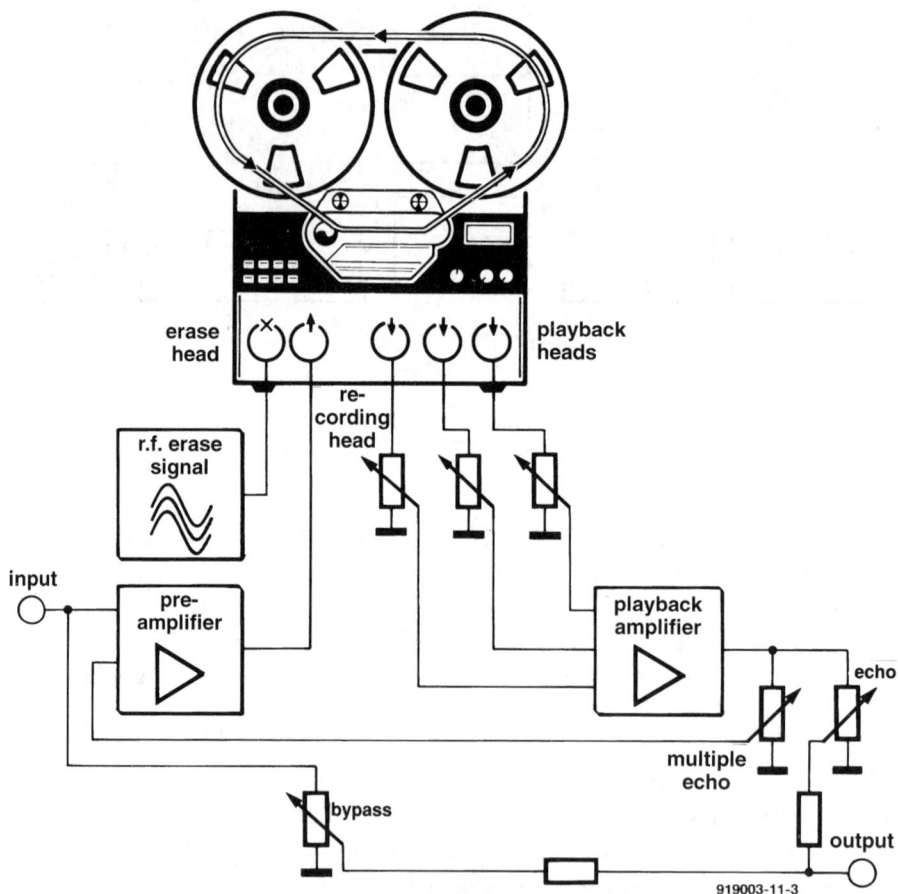

Figure 5.1.5. Principle of a tape recorder delay unit. The signal is recorded on to an endless tape. Playback is effected by three heads, giving rise to three different echoes. The level of each echo can be varied with a preset potentiometer. The output signal is recorded anew so that a very large number of echoes can be obtained. The original input signal is mixed with the output signal via the bypass control.

$$\varDelta t = d/v,$$ [Eq. 42]

where d is the spacing in cm between the recording head and the playback head, and v is the tape speed in cm/s.

230

The tape speed can be varied continuously only in some expensive tape echo machines. Most others only provide a few steps of adjustment. The use of a number of playback heads allows different echoes to be mixed in with the signal. A snag with these machines is their maintenance (they are largely mechanical devices), and the life of heads and tape is limited. The endless tape needs to be replaced fairly regularly particularly when it is short.

The remarks made earlier about digital reverb units apply basically also to digital delay lines. The signal is sampled, quantized, delayed, reconverted and finally mixed with the original signal. A digital echo unit is much easier to design than a digital reverb unit, which basically consists of a large number of echo units. Many digital delay lines have facilities for producing reverberation. The time delay can usually be set in tiny steps of, say, 1 ms. Good digital delay circuits meet the most stringent requirements and can be used over the entire audio frequency range.

In a digital effects unit, the effect is determined by the computer program. There are programs for other artificial effects than those discussed, such as flanging, phasing, harmonizing and chorus.

5.1.3 Linear distortion

Linear distortion changes the amplitude but not the shape of the audio signal. Circuits that provide linear distortion are generally known as tone control circuits.

Standard types of tone control for bass, treble and midrange frequency control were discussed in Chapter 3, but in this section the more complex tone controls, the so-called equalizers, will be discussed. There are three sorts of equalizer: graphic, parametric and paragraphic.

A graphic equalizer enables the signal amplitude within fixed frequency bands to be varied. There are normally up to 30 of these bands, although 10, each an octave wide, is usually quite sufficient. The signal level in each of the bands is set with slide potentiometers, which are clearly seen in the photograph in Figure 5.1.6. This arrangement provides a good overview of the set frequency response. When all controls are in their mid-position, the equalizer

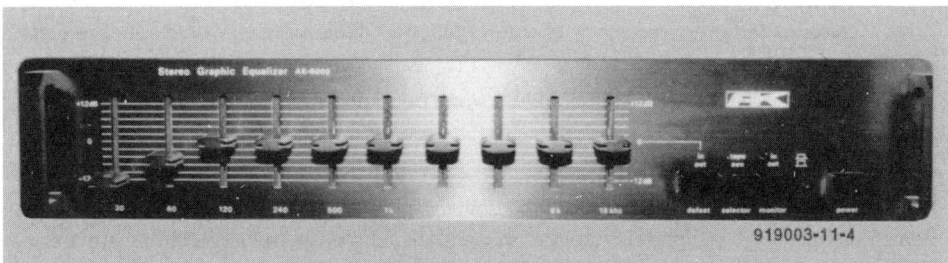

919003-11-4

Figure 5.1.6. 10-band graphic equalizer for hi-fi applications.

has no effect on the signal. Shifting a control up (amplification) or down (attenuation) varies the tone of the audio signal.

There are vast differences in quality between the many types and models on the market. Apart from low noise, it is important that the response of the unit is straight when all slide potentiometers are in the same position. When, for instance, each potentiometer is set to amplify the signal in the relevant band by 3 dB, the overall response should be straight. Unfortunately, this is the case in only a handful of equalizers. Normally, the response is lumpy as soon as the controls are shifted from their mid-position—see Figure 5.1.7.

Since the frequency bands cannot be shifted by the user as required, for instance, to sup-

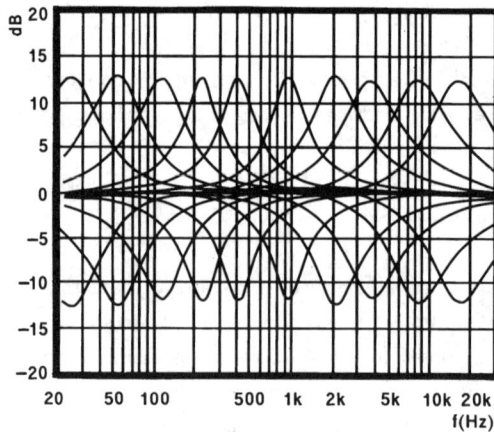

Figure 5.1.7. Response curves of a 10-band graphic equalizer. These curves show
the response as a function of the position of the slide potentiometers.
The mid-frequencies can be attenuated or amplified by ±13 dB with respect to the input
level. When all potentiometers are in their mid-positions, the response is not straight.

press a certain interfering frequency, and owing to their dimensions, graphic equalizers are not suitable for incorporating in a mixer.

Parametric equalizers, however, enable a small band around a certain frequency f_o to be amplified or attenuated. Parameters that can be set are: the central frequency, f_o, the degree of amplification or attenuation, and, sometimes, the width of the frequency band. This is shown in a simple way in Figure 5.1.8. So, a parametric equalizer has at least two or three potentiometers, but the set positions cannot be overseen in one glance. In fact, when several parametric equalizers are used side by side, it is quite difficult to assess the settings.

232

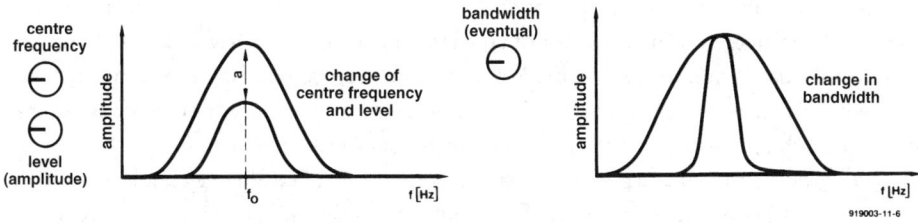

Figure 5.1.8. Operation of a parametric equalizer. The central frequency, f_o, the amplitude, and the bandwidth can all be set independently.

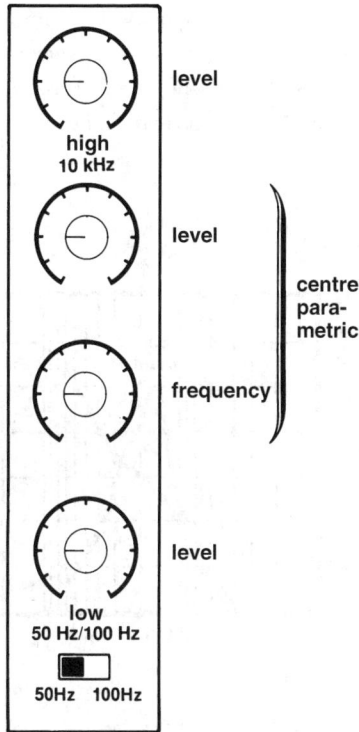

Figure 5.1.9. Control facilities of a paragraphic equalizer.

Nevertheless, the parametric equalizer has the enormous advantage of being highly flexible since the frequency to be raised or attenuated can be chosen freely over a wide range. Moreover, the unit is small enough to make it suitable for incorporating in a mixer. In large mixers, there are often two or three parametric equalizers for each input channel: one for the bass, one for the treble, and the third for the midrange frequencies.

A paragraphic equalizer is a cross between a graphic and a parametric equalizer, which is frequently used in mixers. The controls of a paragraphic equalizer are shown in Figure 5.1.9. The central frequency for the treble range is fixed at 10 kHz, that for the bass frequencies can be switched between 50 Hz and 100 Hz, but that for the midrange is fully parametric, that is, the central frequency and the degree of amplification or attenuation can be varied at will.

The wah-wah pedal (often called cry-baby pedal) also provides linear distortion. In fact, it is a kind of parametric equalizer whose central frequency can be shifted by a foot-pedal. The frequency range over which it can be shifted is tuned to an electric guitar and is about 200 Hz to 2 kHz. The frequency response of such a pedal is shown in Figure 5.1.10. Note that this peaks at about 100 Hz when the pedal is up, and about 800 Hz when it is fully depressed, and that the width of the characteristic changes with the position of the pedal.

Apart from foot-operated wah-wahs, there are versions in which the centre frequency is shifted automatically over the whole frequency range. This gives the guitarist more freedom of movement, but the play and sound variations are limited.

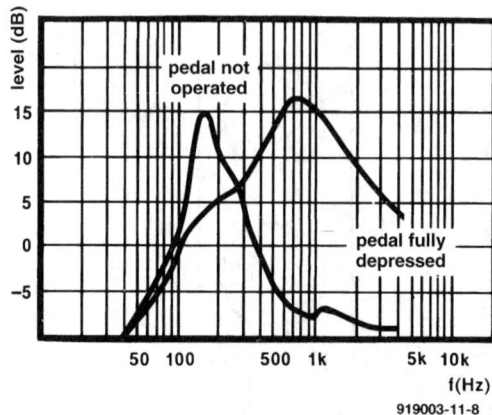

Figure 5.1.10. Frequency response of a wah-wah pedal intended for use with an electric guitar. In the illustration, not only the central frequency but also the width of the characteristics depends on the position of the pedal.

5.1.4 Non-linear distortion

When non-linear distortion is introduced, the shape of the original input signal is altered. Equipment producing non-linear distortion introduces additional vibrations (overtones, harmonics, partials) which were not present in the original signal. The frequency of an overtone is always a whole multiple of the fundamental frequency.

Non-linear distortion is used almost exclusively with electric guitars. The timbre of the distortion unit is determined by the number of harmonics and their amplitude.

As with all music and musical instruments, the final sound is purely a matter of taste. Many guitarists do not use any effects units at all, but provide harmonics in the final sound of their instrument by heavily overdriving the guitar amplifier. For this, a master volume control is an absolute must. Particularly in combination with a valve amplifier, this can result in a very pleasant sound. Moreover, a valve amplifier has a larger dynamic range.

There are many distortion units available, probably because they are relatively easy to design and construct, and because to the user only the subjective sound is of interest. With these units, it is not necessary, as with delay lines and reverb units which simulate natural variations of sound, to take into account any natural rules, only trends in musical taste.

The simplest distortion units transform the guitar output into rectangular waves, which contain a very large number of harmonics. The higher the number of harmonics, and their amplitude, the less is left of the original guitar sound. Distortion units can be used with any type of guitar: as long as the guitar is tuned correctly, they all sound the same; a chord assumes the character of noise.

As with other equipment, before a distortion unit is bought, try it out with your own instrument and amplifier. The technical specification of a distortion unit is totally meaningless, except insofar as the power consumption is concerned (how long will the battery last?).

5.1.5 Level modifiers

Besides equipment that modifies the music signal in a frequency-dependent way, there is apparatus that only affects the signal amplitude: preamplifiers, dynamic compressors and limiters.

When the signal amplitude of a musical instrument is too low, or when very long interconnecting cables are used in the installation, a preamplifier can be inserted in the signal path between the instrument and the amplifier. The amplification of a preamplifier must be constant over a very wide frequency range. The level of the output signal is set with a potentiometer. Tone control is not necessary, since this would only make operation more complicated; in any case, the instrument and amplifier usually have this facility.

A preamplifier may also be used to match output and input impedances (low-impedance output of an instrument to high-impedance input of amplifier or vice versa). Since the preamplifier is intended to magnify the level of weak signals, it is essential that it generates virtually no noise.

The preamplifier for home construction, described in Section 3.3.4 is suitable for this kind of application.

In contrast to that of a preamplifier, the amplification factor of a limiter or compressor is not constant, but depends on the level of the input signal. These units reduce the dynamic range of the input signal in different ways. A compressor compresses the signal: a weak signal is amplified, but a strong one, attenuated. After the signal has passed through a compressor, it has an upper and lower value as shown in Figure 5.1.11, but its frequency and shape remain unaltered. A compressor makes it possible for a guitar tone to be sustained longer, because when the tone becomes softer as the movement of the string decays gradually, this is resisted by the compressor by raising the amplification factor.

Other applications for the compressor are to be found in the recording studio. For instance, it is frequently beneficial for a signal to be compressed before is processed by electronic equipment. Recorded sound on vinyl gramophone records and musicassettes is always compressed since the original dynamic range of the signal cannot be processed by either the sound carrier or the signal-processing electronic circuits.

A limiter limits the signal level once a predetermined threshold has been reached, for instance, to prevent an amplifier being overloaded.

5.2 Cables and leads

The importance of the role of interconnecting cables and leads is usually underestimated. Within a musical installation there are three kinds of cables and leads.

- Mains leads
- Loudspeaker cables
- Signal-carrying leads

Each of these has its own special properties to serve its function satisfactorily. Just as a mains lead is not suitable as a guitar lead, a screened microphone cable is not suitable as a loudspeaker cable.

5.2.1 General

Any electrical cable or lead has a certain resistance, whose value depends on its length, l, its cross-sectional area, A, and the specific resistivity, ρ, of the conductor material. The resistance at room temperature (20 °C), R_{20} is

$$R_{20} = \rho l / A,$$ [Eq. 43]

236

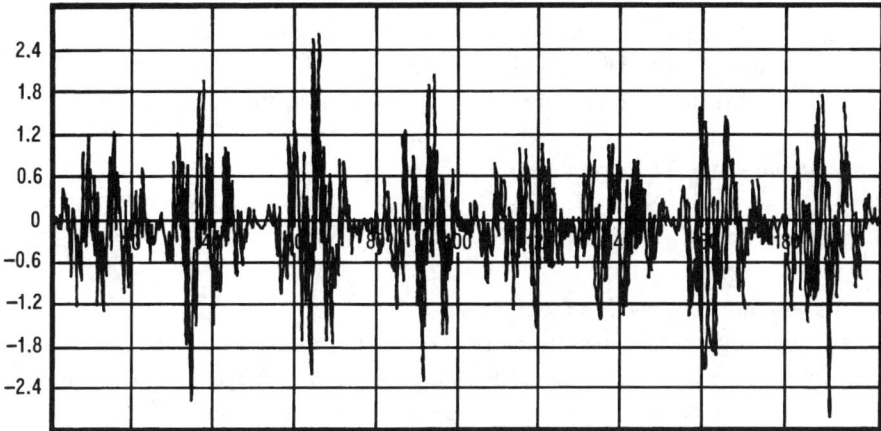

919003-11-9

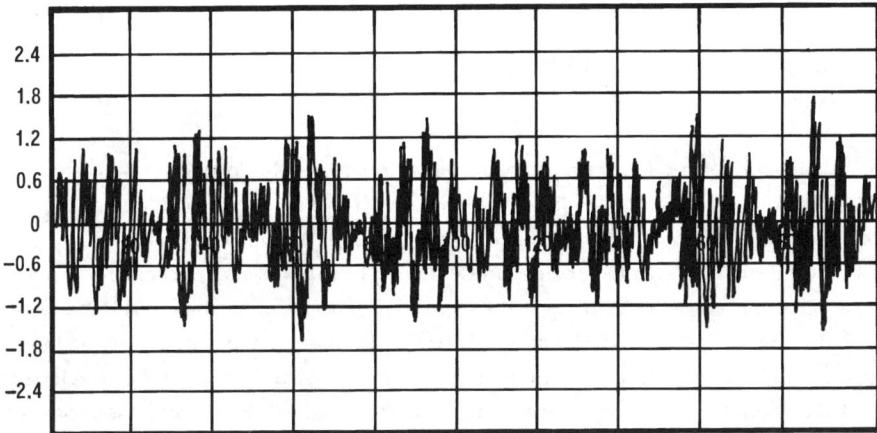

919003-11-10

Figure 5.1.11. Operation of a dynamic compressor. In the upper waveform,
the amplitude of the input signal varies from weak to strong: some signal peaks
exceed a relative value of 2.4. The lower waveform shows that after the signal has passed
through a compressor, its shape is unaltered, but the difference between its lowest and
greatest values is much smaller: the signal peaks do not exceed a relative value of 1.8,
but the lowest levels are appreciably higher than in the original signal.

237

where R_{20} is the total resistance of the lead at 20 °C
            ρ is the specific resistivity of the conductor material in mΩ at 20 °C
            l is the length of the cable in metres
            A is the cross-sectional area of the conductor in square metres

The cross-sectional area of the conductor is

$$A = d^2\pi/4,$$
                                                                      [Eq. 44]

where d is the diameter of the conductor in metres.

Solid copper wire is not very flexible, and that is why stranded wire (hook-up wire or circuit wire) is used. This consists of a number of thin copper wires twisted together. The wire is identified by the number of strands and the diameter of each strand, for instance, 7/0.22 mm means that there are seven strands each with a diameter of 0.22 mm.

The specific resistivity of the conductor material at 20 °C is published in tables; some values of interest are (in Ω)

copper	1.69×10^{-8}
silver	1.51×10^{-8}
gold	2.2×10^{-8}
solder	15×10^{-8}

Obviously, the longer the lead, the greater its resistance, but the larger the diameter, the lower its resistance. The resistance of signal-carrying wires and loudspeaker cables is given in ohms/kilometre (Ω/km). The lower the resistance, the smaller the losses in the cable.

Apart from the ohmic resistance, cables and leads also have parasitic capacitance and inductance, which are evenly distributed over the wire. Their reactance, combined with the resistance of the wire, causes an impedance which may be troublesome in signal-carrying leads. The equivalent electrical circuit of a signal-carrying wire is shown in Figure 5.2.1. The longer the wire, the larger the values of capacitance, inductance, and resistance, and, therefore, the

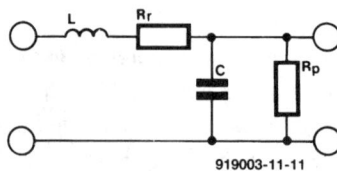

919003-11-11

Figure 5.2.1. The equivalent electrical circuit of a cable.

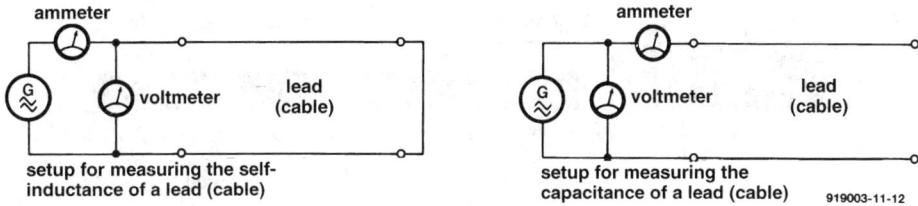

Figure 5.2.2. Test setups for measuring the inductance and capacitance of a cable.

impedance. The inductance and capacitance of a cable can be measured by the test setups shown in Figure 5.2.2, where G is an audio-frequency sine-wave generator. In wire used for signal carrying purposes, the parasitic capacitance is normally specified by the manufacturer in picofarad per metre (pF/m).

The insulation of cables and leads needs no further consideration here as long as mains cables are to BS6500 1990 and insulated hook-up wire is to DEF61-12. Mains plugs (in the UK) should be to BS1363A.

5.2.2 Supply leads

There are two kinds of supply lead of importance in a music installation:

- mains leads for 240 V a.c.
- low-voltage leads

Important parameters of mains cable to BS6500 1990 for various currents are

Max current	3A	6A	13A
conductor size	16/0.2 mm	24/0.2 mm	40/0.2 mm
conductor area	0.5 mm^2	0.75 mm^2	1.25 mm^2
cable dia.	5.6 mm	6.9 mm	7.5 mm

Low-voltage supply wires are for connecting, say, an effects unit to a mains adaptor or battery, where the polarity of the connections is important. An unprotected unit may well suffer serious damage, or even be destroyed, if it is connected to a supply with incorrect polarity. Normally, the enclosure of the equipment (if metal) and the screens of signal-carrying leads are linked to the supply protective earth (in the UK, the large pin at the top of a three-flat-pin plug). If this is not so, as shown in Figure 5.2.3, serious damage (at worst) or blown fuses (at best) ensues.

239

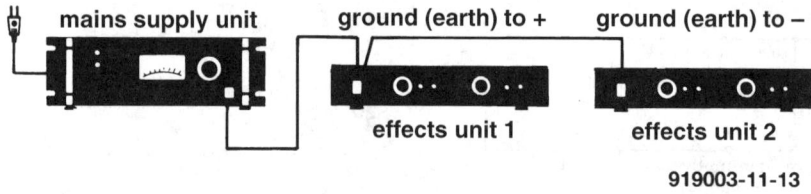

mains supply unit   ground (earth) to +   ground (earth) to –

effects unit 1   effects unit 2

919003-11-13

Figure 5.2.3. If the enclosures of two effects units have dissimilar links to the supply earth, interconnecting them will result in a serious short-circuit.

The insulation of low-voltage wires need not meet stringent requirements, although that of hook-up wire to DEF61-12 has a maximum working voltage of 1000 V, which is, of course, more than adequate.

Low-voltage wires are normally connected to equipment by special low-voltage sockets as shown in Figure 5.2.4. Unfortunately, the polarity of these sockets is not internationally standardized. If in doubt, check it with a multimeter. Sometimes, 3.5 mm mono plugs and sockets are used for low-voltage supply entries, but this is not recommended.

5.2.3 Loudspeaker cables

In the hi-fi world, no cable is the subject of so many misconceptions as loudspeaker cable. This

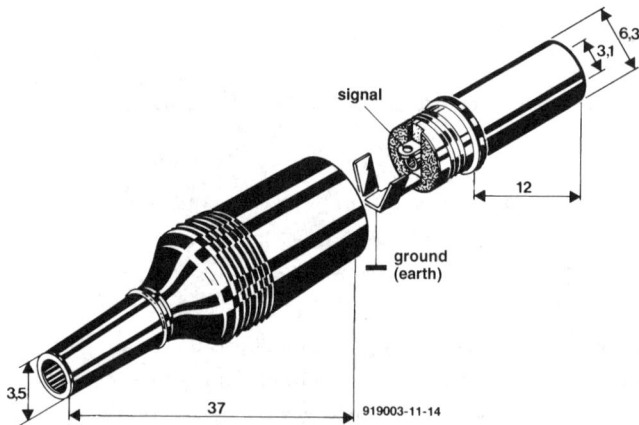

signal

ground
(earth)

919003-11-14

Figure 5.2.4. This type of low-voltage socket has become a de facto industry standard. Unfortunately, its polarity is not standardized internationally.

240

is generally not so in the music world. Since loudspeaker cables usually do not carry high voltage, their insulation is important mainly for its mechanical properties. However, the current through these cables can be high, so the ohmic resistance is an important property. If, for instance, a 4 Ω loudspeaker is to be linked to a 100 W amplifier at a distance of 10 m by a copper cable with a cross-sectional area of 0.75 mm^2, the cable resistance at room temperature is (according to Eq.43):

$$R_{cable} = \rho l / A = 1.69 \times 10^{-8} \times 20 / 0.75 \times 10^{-6} = 0.45 \ \Omega.$$

The voltage across the voice coil of the loudspeaker is

$$U = \sqrt{(PR)} = \sqrt{(100 \times 4)} = 20 \ V$$

The current through the voice coil, ignoring the loudspeaker leads, is

$$I = P/U = 100/20 = 5 \ A.$$

However, with the added resistance of the loudspeaker leads, the current is

$$I = U/R_{LS} + R_{cable} = 20/4.48 = 4.46 \ A.$$

So, the electrical power supplied to the loudspeaker is

$$P = I^2 R_{LS} = 4.46^2 \times 4 = 79.6 \ W.$$

This means that 21.4 W of electrical power is lost in the loudspeaker cable. If the cross-sectional area of the cable had been 2.5 mm^2, the loss would have been only 6.8 W. If, however, a much thinner cable were used, as is often the case, the loss in the cable would be very substantial, indeed. On the other hand, bear in mind that a power loss of 50 per cent would result in a reduction in sound pressure level of 3 dB, which is only just audible.

It is advisable to use XLR connectors to terminate loudspeaker cables; these have a much smaller transfer resistance than 6.3 mm phono connectors. However, even better types than XLR are coming on the market. Loudspeaker cables can easily be made at home (which is considerably cheaper than buying them ready made), but this requires some competence in soldering.

The leads of headphones, which are often used in the recording studio, are normally terminated in 6.3 mm jacks.

The connections to the various connectors are shown in Figure 5.2.5

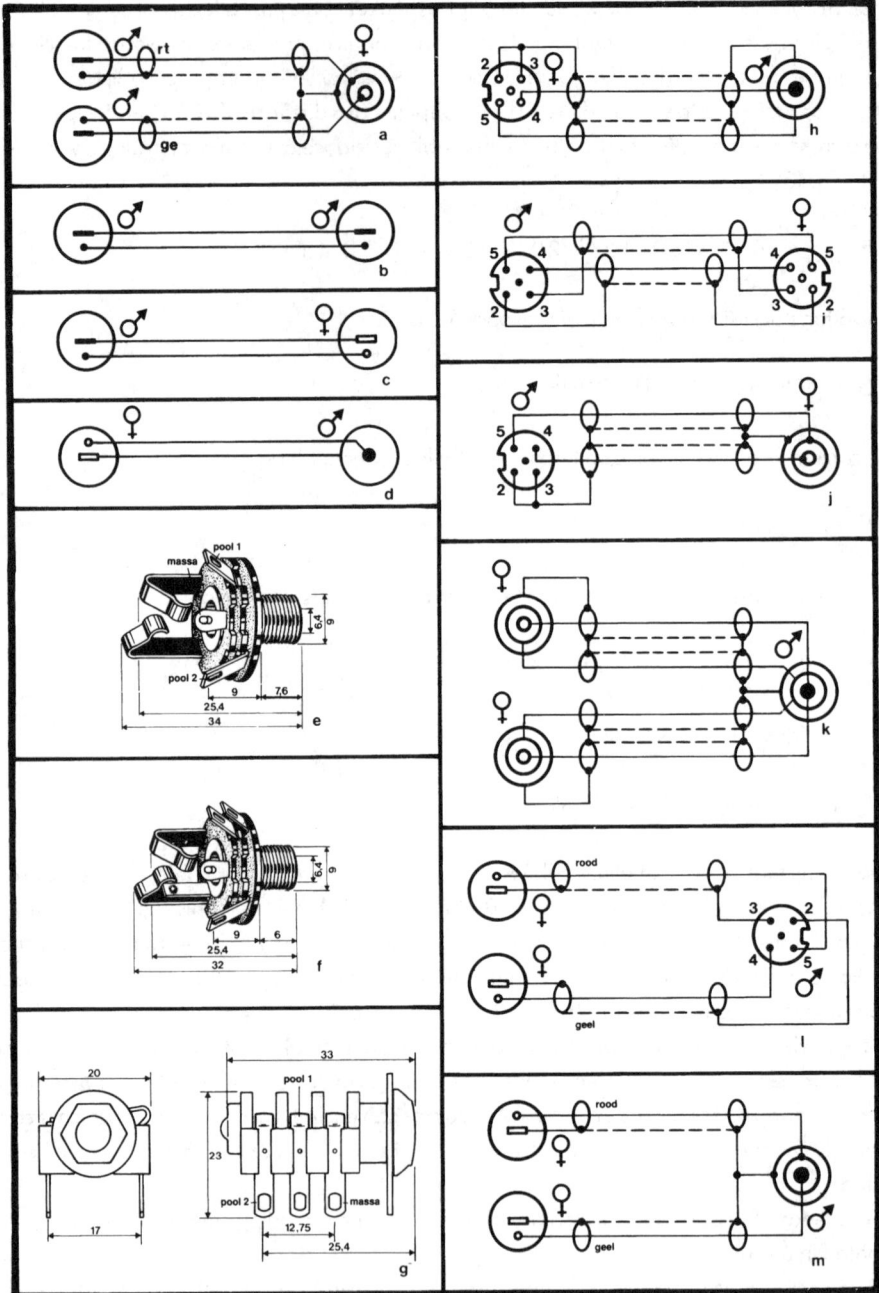

5.2.4 Signal-carrying leads

Interconnecting cables carrying audio signals must always be screened types, irrespective of whether the connection is balanced or not. The quality of the screen determines the effect, if any, of sources of interference on the signal: densly braided (tinned) copper screen is preferred. Loosely braided screens are not satisfactory and do not have a long life.

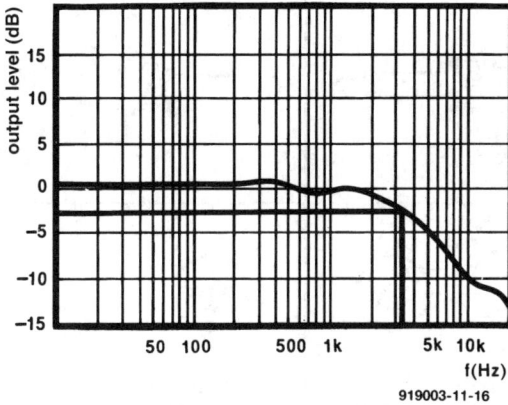

919003-11-16

Figure 5.2.6. Transfer characteristic of a guitar cable.

Figure 5.2.5. (Opposite page) Various connections for loudspeakers and headphones.
a) 2 phono plugs to 6.3 mm socket; 2-core screened lead
b) phono plug to phono plug; 2-core non-screened cable)
c) loudspeaker extension cable; phono plug to phono socket; 2-core unscreened cable
d   phono socket to 3.5 mm plug; 2-core unscreened cable; connection for headphones to 3.5 mm socket
e) 6.3 mm, 3-way socket for chassis mounting
f) same as e) but with switch
g) insulated chassis mount socket
h) headphone cable; 5-pin DIN to 6.3 mm stereo phono socket; 2-core individually screened lead
i) headphone extension cable; 5-pin DIN to 5-pin DIN
j) 5-pin DIN to 6.3 mm stereo phono socket; 2-core individually screened lead
k) splitter lead for connecting, say, two headphones to one amplifier output
l) 2 loudspeaker plugs to standard headphone plug; 2-core individually screened lead
m) 2 loudspeaker plugs to 6.3 mm stereo phono socket; 2-core individually screened lead

243

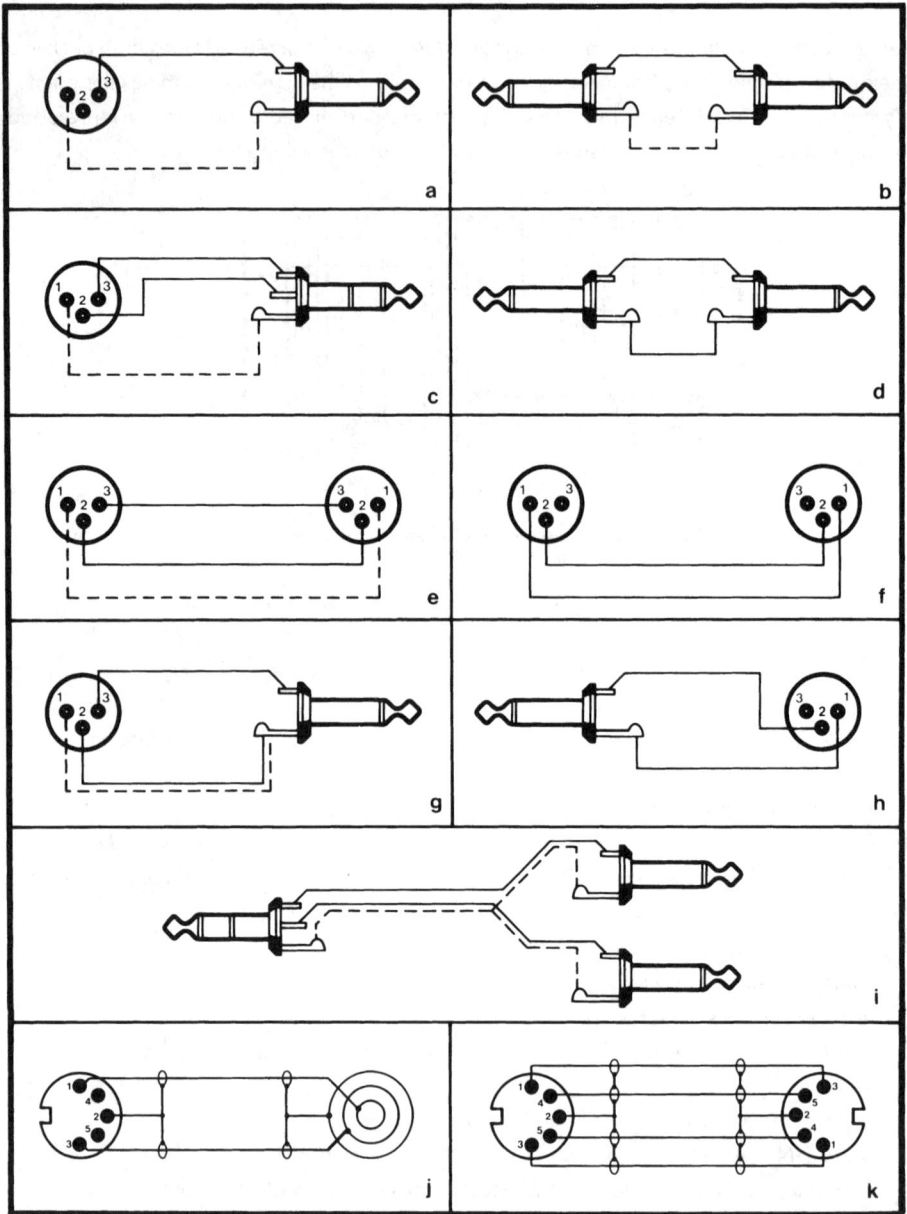

919003-11-17

244

Because of the high input impedance of amplifiers, the ohmic resistance of signal-carrying leads is of not much importance. However, the parasitic capacitance and self-inductance of these cables can prove troublesome. These properties change the cable into a low-pass filter whose cut-off frequency depends on the length of the cable and the capacitance. The cut-off frequency of an eight metre long cable with a capacitance of 250 pF/m and assuming an internal resistance of the signal source (guitar pick up) of 10 kΩ is only 8 kHz. The transfer characteristic of a proprietary spiral guitar cable is shown in Figure 5.2.6. In its non-stretched state this cable is 1.2 m long. Note that the signal level has dropped by 3 dB referred to that at 1000 Hz at a frequency as low as 3000 Hz.

However, the capacitance of the interconnecting cable can have a beneficial effect on the sound of an electric guitar. This is shown by the fact that guitarists who have used a certain cable for years and then convert to a lead-less connection do not like the sound since this then contains too much high. This can, of course, be remedied by the tone control of the amplifier or with the aid of an equalizer.

The connection between instrument and amplifier is usually unbalanced, that is, the cable consists of a single screened wire (the braid is used as the return line). Only few instruments have a balanced output that needs two-core individually screened cable. Since the distance between instrument and amplifier is rarely more than five metres, an unbalanced connection provides adequate protection against hum and noise. Note that if two-core screened cable is used for an unbalanced connection, only one core and the screen should be used to prevent

Figure 5.2.7. (Opposite page) Signal-carrying connecting cables.
a) Link between an unbalanced XLR output and an output amplifier or effects unit with an unbalanced phono plug.
b) Unbalanced link between two units with phono sockets at the output.
c) Link between an insulated, balanced XLR plug and stagebox with stereo phono sockets.
d) Link between amplifier output and loudspeaker with two phono plugs.
e) Link between units with insulated balanced XLR input and output connectors.
f) Link between amplifier output and loudspeaker via XLR connectors.
g) Link between a unit with insulated balanced XLR output and a unit with unbalanced mono socket.
h) Link between amplifier and loudspeaker with a phono plug at one end and an XLR connector at the other.
i) Adaptor connection to split up the two signals of a stereo output.
j) Adaptor cable 5-pin DIN to stereo phono socket; two-core individually screened leads.
k) Stereo connecting cable, 5-pin DIN to 5-pin DIN; four-core individually screened leads; pins 1, 3 and 4, 5 are linked cross-ways.

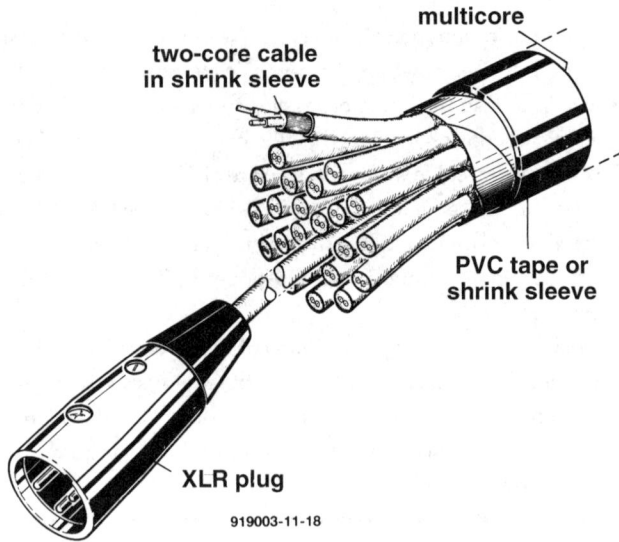

two-core cable
in shrink sleeve

multicore

PVC tape or
shrink sleeve

XLR plug

919003-11-18

Figure 5.2.8. The two-core leads in a multiway cable must be terminated individually.

problems with earth loops or excessive cable capacitance.

Effects units should be interconnected with unbalanced cables about 30 to 60 cm long. Right-angled plugs and sockets are recommended.

DIN connectors are frequently used to terminate signal-carrying leads, and some mixers have DIN connectors. For that reason, a phono jack to DIN adaptor is sometimes required and this can readily be made at home. A number of different connections are shown in Figure 5.2.7.

The connection between a mixer and the stage is normally a multiway cable, which contains a number of two-core leads for balanced connections. The multiway cable is terminated at the stage end in the stagebox, that is, the unit to which all microphones are linked. A multiway cable can contain up to 32 screened 2-core wires and would have a diameter of about 20 mm. The capacitance between the core and braid is usually quite large: of the order of 200–300 pF/m. The effect of this, and that of the resistance, can normally be countered with the tone control and gain control on the mixer.

Before a multiway cable can be used, it has to be terminated, which is normally carried out on the individual two-core leads. Each of these must be clearly marked to prevent time-consuming searches when an installation is being set up. See Figure 5.2.8.

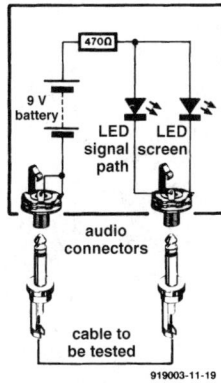

Figure 5.2.9. Simple tester for unbalanced connecting cables terminated into phono plugs. Similar testers may be built for testing balanced connecting cables and cables terminated into different connectors.

5.2.5 Testing cables and leads

The capacitance, C, of a cable is measured with a setup as shown in Figure 5.2.2 and calculated in picofarad from

$$C = 10^{-12}/2\pi fl(U/I),$$

where f is the frequency to which the sine-wave generator is set, l is the length of the cable, U is the reading on the voltmeter, and I is the reading of the ammeter.

Most faults in a cable are breaks in the conductor or braided screen, and these are readily ascertained with a multimeter. Unfortunately, it is not always easy to detect a break by bending and feeling. In that case, cut the cable into two, so that at least one length is perfectly all right for use, say, to connect two effects units. The other half is best discarded.

Electrical testing of cables is made convenient by a simple test circuit as shown in Figure 5.2.9. The example shown is for testing cables terminated into phono jacks, but similar circuits can be built for the testing of cables with other kinds of connector. The tester is powered by a 9 V battery. If the cable is sound, both LEDs light. When it has a break, the LEDs will show whether the break is in the core or in the screen.

6. Recording & reproduction techniques

Transmission embraces the recording, transfer and reproduction of music or vocal performances. In the transmission, the signal must not be altered but arrive at the recipient as faithfully as possible. The distance between the location of recording and that of reception of the reproduced sound is immaterial. In radio, large distances are involved; in a concert, distances of metres only: from stage to audience. Therefore, as far as sound engineering is concerned, a public-address system is as much part of the transmission system as the recording equipment in a studio.

For faithful reproduction, it is essential that the entire audio frequency range is processed undistorted by the system. This is, of course, true of all links in the reproduction chain. Clearly, faithful reproduction is attainable only with extensive technical tools and methods. In contrast to musicians' amplifiers and mixers, in the transmission we are concerned only with figures and objective facts. The concept of preference is immaterial since all that is required is the faithful reproduction of the music as it is played on the stage. Of course, this is only partly possible; particularly where dynamic range is concerned: losses must be allowed for.

6.1 Hi-fi, mono and stereo

The term hi-fi (high fidelity) is used with reference to transmission equipment that meets a number of special requirements. There requirements are laid down in DIN45500. Some of the more important ones are low distortion and large frequency range. The term hi-fi has nothing whatever to do with the designations mono(phonic) and stereo(phonic), but merely refers to the transmission quality of a reproduction channel.

People have two ears which enables them to hear spatially. This means that we can hear differences in path time and sound pressure that arise from reflections of sound waves by walls, ceilings, and furniture around us. When a recording is made with just one microphone, this information is lost irretrievably since a microphone can pick up sound in one location only. This is called mono(phonic) transmission, which is perfectly acceptable for speech.

When the spatial impression and the positional information of individual instruments must be retained, two-channel transmission must be used. All constituent parts of the transmission chain are then required in twos: microphones, amplifiers, loudspeakers, and so on. In this case, we speak of stereophonic transmission. It makes the sound transparent. The reproduction is particularly faithful if during the recording a dummy head (kunstkopf) is used. This is a replica of a human head in which the eardrums have been replaced by small microphones, so that the sound is picked up in the same way as we hear it.

6.1.1 Stereophonic reproduction

When a monophonic signal is reproduced at the same level by two identical but well separated loudspeakers, we get the impression that we hear only one source of sound somewhere in the middle between the two loudspeakers. When the sound level to the loudspeakers is made different, the virtual location of the sound source moves correspondingly. If, say, the sound d pressure level to the left-hand loudspeaker is 6 dB higher than that to the right-hand loudspeaker, the sound source appears to move to the left, about halfway between the left-hand loudspeaker and the centre between the two loudspeakers. When the difference in SPL is made 12 dB, the sound appears to come from very near to the left-hand loudspeaker. At greater level differences, the sound appears to come from the left-hand loudspeaker only. These situations assume that the listener is positioned on a line midway between the two loudspeakers and not too close to them and are represented graphically in Figure 6.1.1. Our ears react to the difference in sound pressure level only. This phenomenon is the basis of stereophonic sound.

Figure 6.1.2 shows the mutual placing of loudspeakers and listeners for optimum stereophonic reproduction. When the listener is close to the loudspeakers, it is important that he/she is exactly on the line midway between the loudspeakers. Further away from the loudspeakers, the listener has more room for moving about. When the distance between loudspeakers and

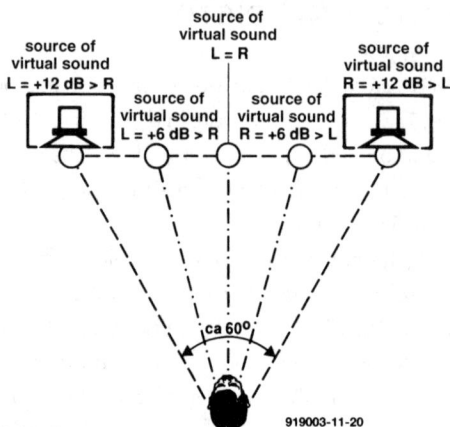

Figure 6.1.1. Illustration of how the virtual source of sound moves
with different power inputs to the two loudspeakers.

250

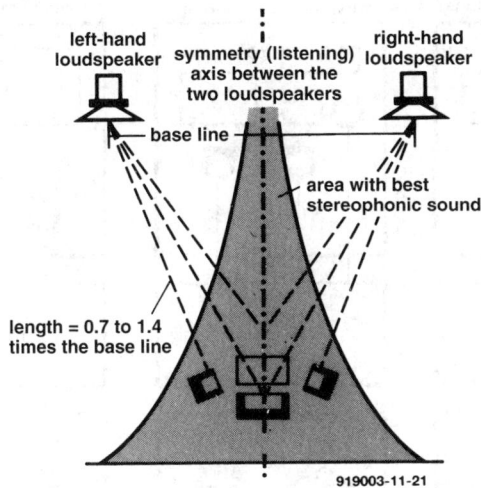

Figure 6.1.2. Optimum stereo reproduction is available only in the shaded area.

listener becomes too large, the stereophonic effect gets lost.

Apart from differences in sound pressure levels, the stereo effect can also be obtained on the basis of path time differences. Such differences contribute little to the virtual location of a source of sound, but they enhance the spatial impression. At frequencies below 300 Hz, a sound no longer has a virtual position. This means that low frequencies may be recorded and reproduced monophonically without any detriment to the stereo image.

6.1.2 Stereophonic recording and transmission

There are three main types of recording and transmission methods:

- by the X–Y or M–S system (coincident microphones);
- with separate microphones;
- by the A–B system.

the difference between them are represented in Figure 6.1.3. Stereophonic methods using the X–Y or M–S system or separate microphones are based on differences in sound intensity. As mentioned earlier, these are fine for localizing the source of sound. During recording, two directional microphones are used at a single location, positioned above one another in such a way that they can be turned in the horizontal plane with respect to each other.

251

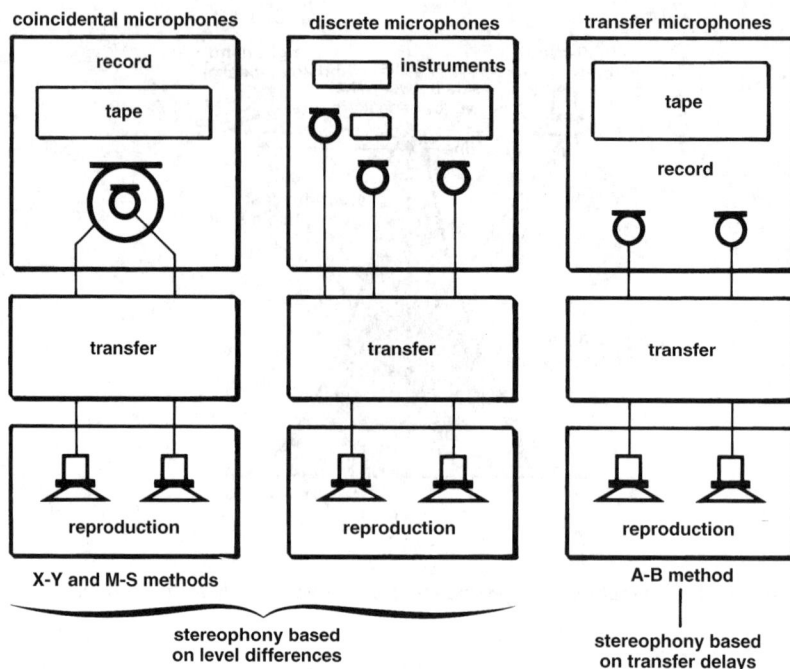

Figure 6.1.3. Different methods for making stereophonic recordings.
In public-address technology separate microphones are used.

Because of the directivity of the microphones, they both provide a signal that depends on the intensity of the picked-up sound. This is called the X–Y method, which is typified by good spatial reproduction and ready location of the sound sources. Also, the acoustics of the recording room is well represented. This is why the method is used frequently for the recording of pop and rock groups, as well as small orchestras (and also for live performances).

The M–S (middle and side) method is a variation of the X–Y method. With this method, a monophonic signal and a difference signal are recorded with which it is possible to reproduce both monophonic and stereophonic sound at the receiving end. The method is used frequently in radio broadcasting. The X–Y and M–S methods are often grouped together and called the coincident method.

The system using separate microphones is of particular interest to bands. In it, each instrument is recorded with a separate microphone. The signals from all these microphones are combined in a mixer. This may be compared with the reproduction via a public-address system. The method is used widely in recording studios particularly for pop and rock groups. The location of the instruments in the stereo image is determined with panpots (panoramic potentiometers).

252

It is therefore possible to relocate the position of the instruments to personal taste, or to do so after the recording has been made (as happens in the studio). Each instrument can be placed at any desired location, which provides a very clear presence. Since the microphones are placed directly in front of the instruments or loudspeakers, a good signal-to-noise ratio is guaranteed. Unfortunately, the acoustics of the recording hall is completely lost; in the studio this is added artificially with the aid of reverb and echo units.

With the A–B or spaced microphone system, two microphones are used at some distance from each other. This results in both intensity and path time differences to be recorded. In rooms with much reverberation, the path time differences are faint. So, even in this case, intensity level differences are the most important. Depending on the positions of the microphones, certain sound sources can be emphasized. The method enables good, spatial recordings to be obtained. It is particularly suitable for live recordings and recordings in the rehearsal room, since the positioning of the microphones is not particularly critical. Consciously or not, most recordings are made with this method. For good results, it is advisable to place the instruments in a straight line in front of, or in a semicircle around the microphones. The voice loudspeakers may be moved slightly forward to ensure that the vocal parts are not (partly) drowned.

6.2 Monitor (hi-fi) amplifiers

Although amplifiers have been discussed in some detail in Chapter 3, hi-fi amplifiers were not included. Yet, they are also used in sound engineering, for example, in a recording studio as monitor amplifier to keep a check on tape recordings. Normally, these are very good quality units to ensure that the reproduction is as pure and faithful as possible. Their frequency range should extend into the ultrasonic region to prevent irregularities with the audible frequencies. The noise figure and distortion must be very small. Since these amplifiers are permanently used in a dry atmosphere, protection circuits and high power ratings are not very important. What is important, is the quality of the reproduced sound. Since the sound d is judged objectively, there is no need for tone control. If this is fitted all the same, the position at which the frequency characteristic is straight must be clearly indicated.

Some of the requirements of hi-fi norm DIN45500 are:
- minimum continuous output power (sine wave):
    mono: 10 W
    stereo 6 W per channel
- distortion factor:
    not more than 1% at the nominal power output, below that, not more than −26 dB
    within a frequency range of not less than 40 Hz to 12500 Hz
- frequency range:
    not less than 40 Hz to 16 kHz ±1.5 dB referred to 1000 Hz.

6.3 Public-address (PA) installations

Public-address installations are intended for sound reproduction in large halls during live performances to enable the audience to hear clearly what happens on the stage. Voice amplifiers were already discussed in Section 3.2. There is no sharp dividing line between these and PA

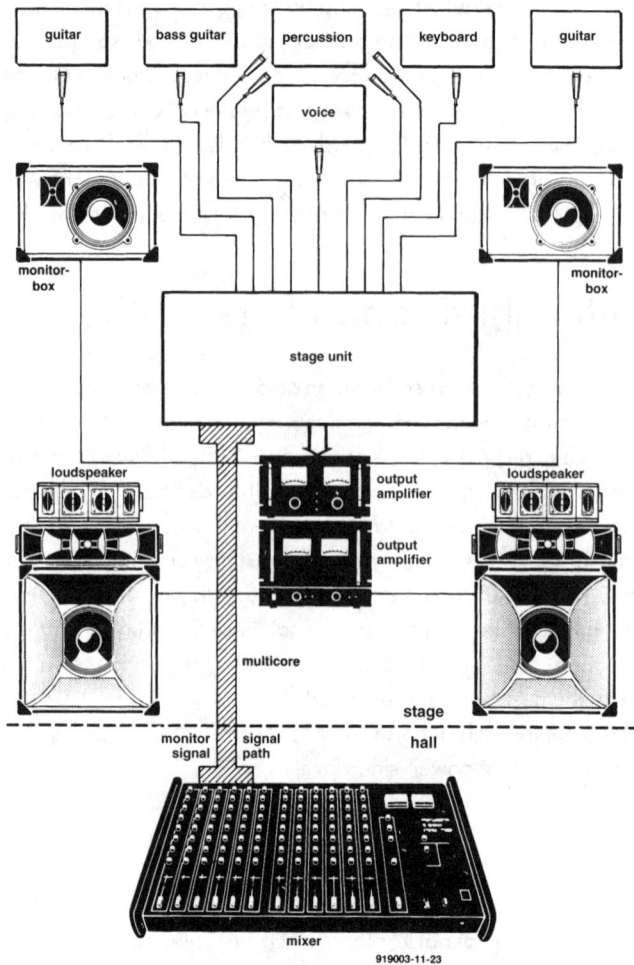

Figure 6.3.1. Public-address installation for small halls.

amplifiers. In other words, the larger and more extensive a vocal installation, the more it resembles a PA system. Apart from the power mixers, the discrete constituent parts differ only as far as dimensions and output power are concerned.

All instruments must be recorded, processed and reproduced with the aid of microphones, mixer, amplifiers and loudspeakers. In addition, the musicians on the stage should have a monitor installation. All in all, this is a very extensive setup. The electronics for it cannot be housed in a single enclosure and this is why a PA installation is always composed of discrete units as shown in Figures 6.3.1 and 6.3.2.

6.3.1 Mixers

A mixer for use in a PA installation (see Figure 6.3.2) differs hardly from that used in a recording studio. The requirements of both are the same, but the signal of a mixer during a live performance is applied to an output amplifier and not to a tape recorder.

Some instruments, such as synthesizers, may be recorded direct from the mixer. This has some advantages, because it reduces the number of microphones on the stage, which reduces the risk of positive feedback, and the background noise is reduced. Moreover, a direct connection is less expensive than a microphone. If, however, the loudspeaker influences the sound of the instrument, direct recording is not possible.

Since, owing to the large number of units involved, it is not easy to get an overall view of a public-address installation, it is essential that with direct connections certain safety aspects are not overlooked. It could happen, for example, that when one of the many units is faulty, there is mains on the enclosure. This means that the entire installation, including any directly connected instruments, may be at mains potential. Also, earth loops may occur between mains

919003-13-9

Figure 6.3.2. Mixer for use in a PA installation or recording studio.

left-hand

10 folded-horn loudspeakers each containing two 38 cm bass drivers

10 mid-high loudspeakers each containing four 25 cm drivers, two 5 cm drivers and four tweeters

output amplifier racks each containing a BGW250D high/mid-high driver, a BGW750C mid-range driver, and a Camco LA801 bass driver

| bass | mid-high | output amplifier rack |
| bass | mid-high | |

| bass | mid-high | output amplifier rack |
| bass | mid-high | |

| bass | mid-high | output amplifier rack |
| bass | mid-high | |

| bass | mid-high | output amplifier rack |
| bass | mid-high | |

| bass | mid-high | output amplifier rack |
| bass | mid-high | |

1 — 27-band equalizer — active crossover filter — output amplifiers

2 — 27-band equalizer — active crossover filter — output amplifiers

3 — 27-band equalizer — active crossover filter — output amplifiers

4 — 27-band equalizer — active crossover filter — output amplifiers

5 — 27-band equalizer — active crossover filter — output amplifiers

divider unit (stage to main mixer and stage to monitor mixer)

microphone and line outputs from the stage

4 4
percussion

bass guitar
3

1
guitar   singer
2   5

A B
C D

A = bass
B = mid-high
C = mid-low
D = high

active four-way crossover filter

27-band equalizer

left-hand   master outputs

multi-core inputs 1–40

mixer with 40 inputs, eight sub-groups and two outputs

Insert
out   in

insert rack with equalizers, compressors and noise gates

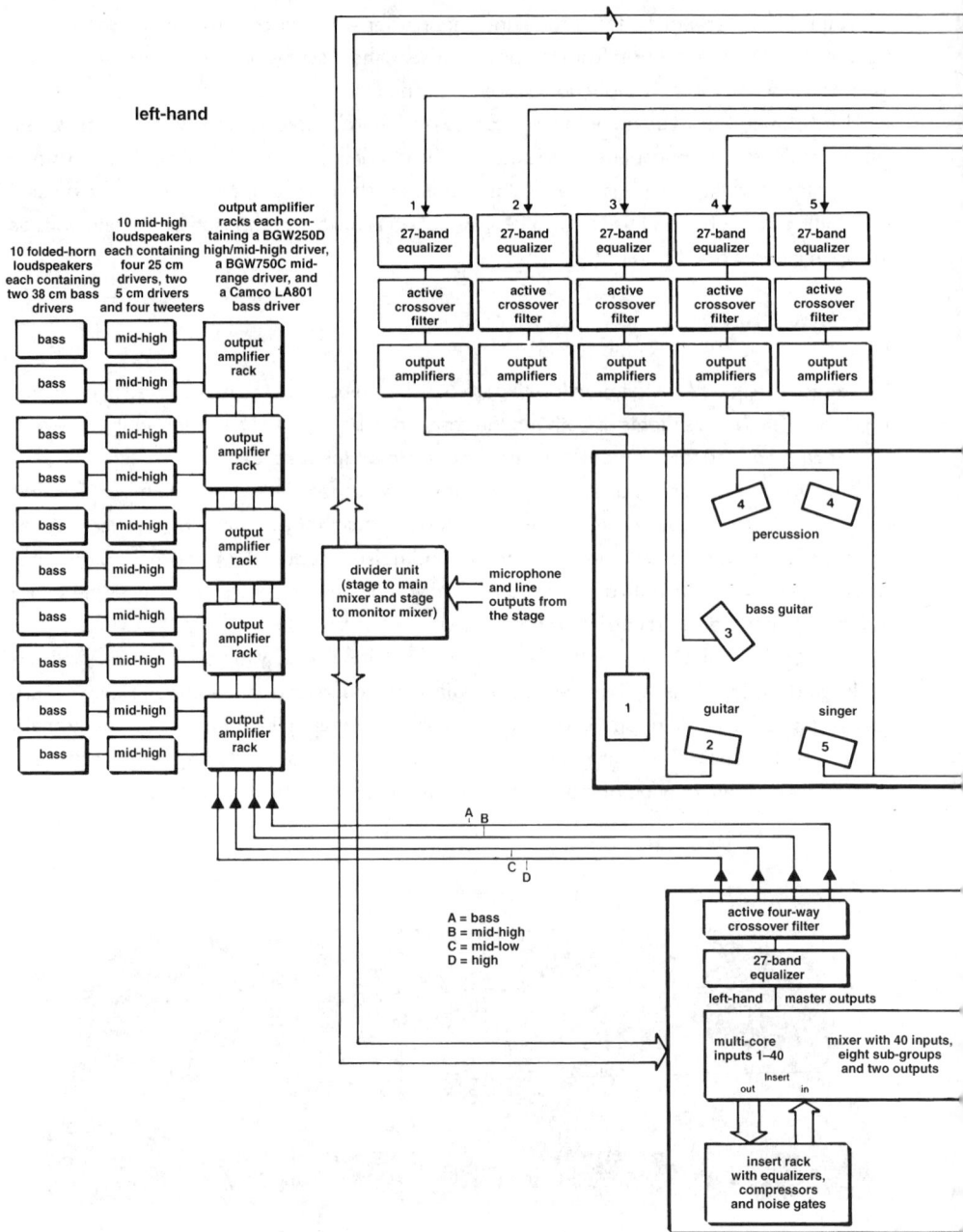

Figure 6.3.3. Typical public-address installa

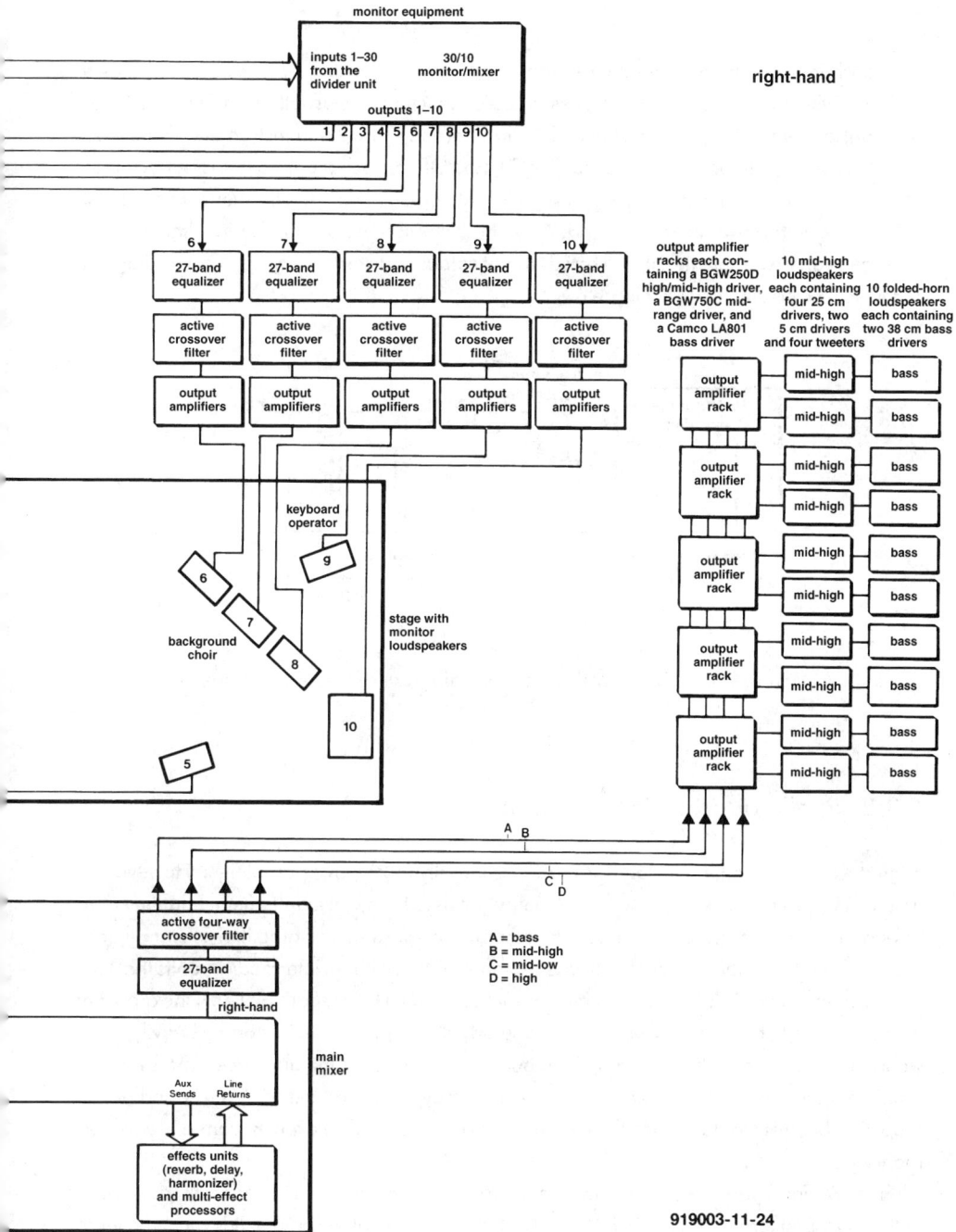

monitor equipment

inputs 1–30 from the divider unit

30/10 monitor/mixer

outputs 1–10

| 1 | 2 | 3 | 4 | 5 | 6 | 7 | 8 | 9 | 10 |

6 7 8 9 10

| 27-band equalizer | 27-band equalizer | 27-band equalizer | 27-band equalizer | 27-band equalizer |

| active crossover filter | active crossover filter | active crossover filter | active crossover filter | active crossover filter |

| output amplifiers | output amplifiers | output amplifiers | output amplifiers | output amplifiers |

output amplifier racks each containing a BGW250D high/mid-high driver, a BGW750C mid-range driver, and a Camco LA801 bass driver

10 mid-high loudspeakers each containing four 25 cm drivers, two 5 cm drivers and four tweeters

10 folded-horn loudspeakers each containing two 38 cm bass drivers

output amplifier rack	mid-high	bass
	mid-high	bass
output amplifier rack	mid-high	bass
	mid-high	bass
output amplifier rack	mid-high	bass
	mid-high	bass
output amplifier rack	mid-high	bass
	mid-high	bass
output amplifier rack	mid-high	bass
	mid-high	bass

keyboard operator

9

6

7

background choir

8

stage with monitor loudspeakers

10

5

A B

C D

active four-way crossover filter

27-band equalizer

right-hand

A = bass
B = mid-high
C = mid-low
D = high

main mixer

Aux Sends   Line Returns

effects units (reverb, delay, harmonizer) and multi-effect processors

919003-11-24

arge halls.

257

earth and signal earth, because the relevant instrument on the stage is often quite a distance away from the mixer. To prevent such risks, direct connections are usually made via a DI (Direct Injection) box which provides electrical isolation between mixer and instrument. Dangerous potentials and earth loops do not occur in a DI box. Often, such a box contains a special isolating transformer, while others use an amplifier with optoisolator. The frequency range must, of course, cover the whole audio band. Also, the distortion factor must be negligibly small. Only then can we be sure that the DI box does not colour or distort the signal. A circuit diagram of a typical DI box is shown in Figure 6.3.4.

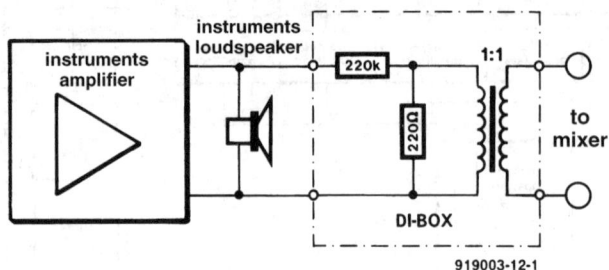

Figure 6.3.4. Circuit and connection diagram of a typical DI box.

6.3.2 Public-address (PA) amplifiers

All that was described in Section 3.2.6 applies equally to PA power amplifiers. However, PA amplifiers have to provide much greater output powers than voice amplifiers, and they must have a much greater dynamic range than common-or-garden hi-fi amplifiers. After all, they must amplify the various musical instruments, that is, direct without intermediary units that limit the dynamic range. This implies that a PA amplifier needs a lot of spare power. If the amplifier output were just about adequate, it would regularly start clipping (i.e., distorting heavily) . This would not necessarily affect the amplifier, but it certainly would the audience. To prevent the amplifier being overloaded, special compressors may be placed between mixer and output amplifier, but this would reduce the dynamic range the audience expects from a live performance.

Since PA amplifiers are usually a fair distance away from the mixer, but normally quite close to the loudspeakers, extensive VU (Visual Unit) meters are not necessary. It is much better to use a fair-sized 'overload LED' that is visible from a distance to indicate that the amplifier is clipping. When the output amplifier has a modest power rating, it is advisable to keep the vol-

258

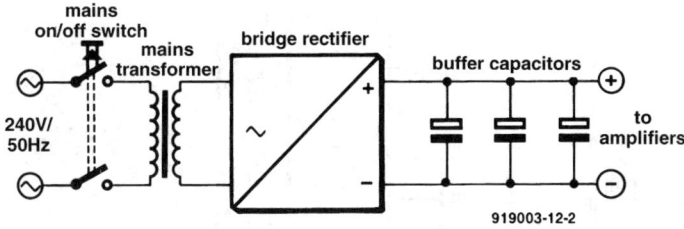

919003-12-2

Figure 6.3.5. Basic power supply for a PA amplifier. The reservoir (buffer) capacitors are discharged on switch-on and therefore draw a very large (surge) current.

ume setting not too high to retain a good dynamic range and to prevent clipping. After all, a performance at a slightly subdued volume but providing good-quality sound is much to be preferred over a distorted din.

The power output may be increased by connecting several output amplifiers in parallel but, just as with loudspeakers, care must be taken to ensure that their input sensitivities are roughly the same (see also Section 3.2). It should be noted that doubling the sound pressure level requires raising the amplifier output ×10. Doubling the output power increases the sound pressure level by 3 dB which is only just audible. It is, therefore, of no practical consequence whether an output amplifier is rated at 500 W or 600 W.

The technical specification of public-address amplifiers must be first class: distortion must be small and the frequency range large – preferably extending to well into the ultrasonic region. Most PA amplifiers on the market meet these requirements.

It is important to take measures that increase the reliability of PA amplifiers, such as those already described in Section 3.2.6: thermal protection; protection against direct voltage and current; and power-on delay. What is equally important is a good power supply. The power supplies of stereo output amplifiers must be electrically isolated. This increases the peak loading of an amplifier and, in case of a fault in one channel, there is a second channel.

The switch-on current of a PA amplifier is large and may cause fast-acting fuses to melt and

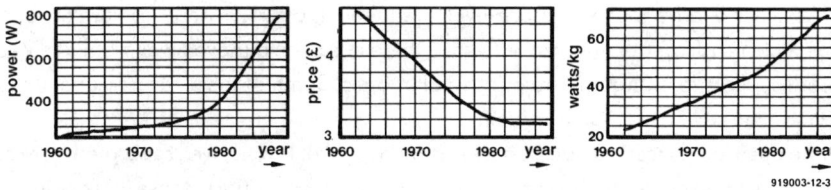

919003-12-3

Figure 6.3.6. The trend in output amplifiers is clearly in the direction of ever greater output power, lower weight and lower prices.

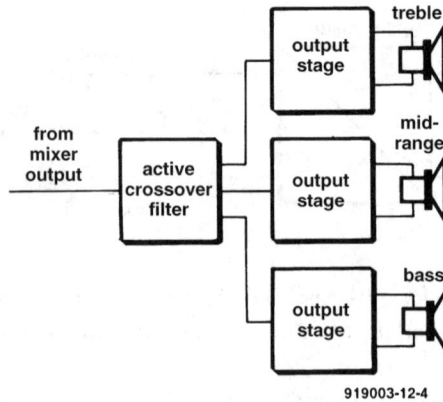

919003-12-4

Figure 6.3.7. In large PA installations, an active cross-over filter divides the frequency range into a low-, middle-, and high-frequency band.

so interrupt the power supply. This large surge current is caused mainly by the reservoir capacitors in the power supply (see Figure 6.3.5). When the supply is switched on, these capacitors are discharged and form a near-short-circuit so that they draw a very large current. This is why many amplifiers have a soft-start facility, such as a daisy-chain power sequencer.

During the history of power amplifiers, the power rating has steadily increased, whereas the dimensions, weight and price have dropped. See Figure 6.3.6.

In large public-address installations, active cross-over filters are used which divide the frequency range into a bass range, mid-frequency range, and treble range. A separate output amplifier is used for each of these ranges (see Figure 6.3.7).

6.3.3 Public-address loudspeakers

Although the loudspeakers in a PA installation must, like all other constituent parts, reproduce the entire audio frequency range as faithfully as possible, they are the weakest link in the chain of PA units. Since a large sound pressure level is required, the current trend is to use exponential loudspeakers. Also, over the past few years, there has been a move away from a few very large loudspeakers to a larger number of smaller ones. Some of these are frequently suspended from the ceiling of a listening hall.

At large open-air concerts a number of large loudspeaker towers, well separated from one another, are used. To prevent echoes between these towers, the signal to those that are furthest away from the stage is slightly delayed by a suitable delay line. The delay time is equal to the time the sound normally takes to travel from the stage to the relevant loudspeakers – usually

260

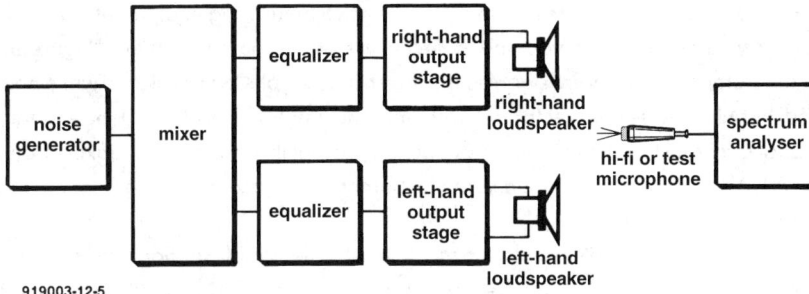

919003-12-5

Figure 6.3.8. Interfacing a public-address installation to the acoustics of a listening or recording area. A spectrum analyser is used to determine any irregularities caused by the acoustics. Equalizers correct these in the equipment and so ensure that the frequency characteristics are brought back to their original shape.

of the order of a few milliseconds (ms).

6.3.4 Interfacing with hall or studio acoustics

A large sound reproduction system must be set up to interface correctly with the acoustics of the listening or recording area, otherwise there is a risk that a live performance could become

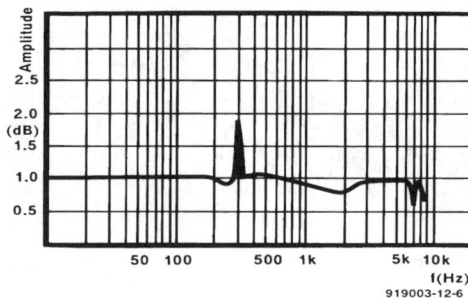

919003-12-6

Figure 6.3.9. Irregularities in the frequency response caused by the acoustics of a listening or recording area, measured with a spectrum analyser in the frequency range 50–8000 Hz. There is a clear resonance peak at about 300 Hz, which can be largely suppressed by an equalizer. The frequencies between 500 Hz and 5000 Hz are slightly attenuated, so that they must be raised accordingly by an equalizer. There are dips at about 6000 Hz and 8000 Hz, which must also be corrected.

261

a distorted din. To do so requires a range of expensive measuring equipment and tools. One way of carrying out tests to determine the acoustics of the room is shown in Figure 6.3.8. A noise generator applies a noise signal to the mixer. In principle, both pink and white noise may be used, but white noise is easier to work with since the amplitude of all frequencies is the same. The signal level must not be too high, however, to prevent the treble loudspeakers from being damaged. Also, many modern measuring instruments take account of the frequency-dependent amplitude of pink noise.

All tone controls in the PA installation must be set to their mid-position. The noise radiated in the area by the loudspeakers is then measured with a standard microphone (with straight frequency response) at a number of locations in the listening area and applied to the spectrum analyser. This instrument indicates how, and to what extent, the frequency response of the PA installation is affected by the room acoustics. Some frequencies may be amplified, while others may be attenuated (see Figure 6.3.9). Equalizers are used to attempt to counter the changes in the frequency response caused by the room acoustics. In an ideal case, the equalizers and the spectrum analyser have similar filter responses.

Often, there are not the means for carrying out such tests (lack of money and skilled personnel), and the interfacing has to be attempted on the basis of listening tests. For this purpose, a recording is played back of which it is known how this should sound, and by listening to it any resonance peaks or absorption dips are determined. Equalizers are again used to try to counter such peaks and dips.

7. Lighting

Apart from sound, light effects are an indispensable part of rock and pop concerts. As far as the lighting system is concerned several points must be borne in mind:

- high energy consumption;
- noise and interference that may affect the sound installation.

Also, it should be borne in mind at all times that a lighting installation is a potential source of risks and dangers. In contrast to the sound installation, high-voltage wires run all the way from the control panel to the lamps. Making such an installation yourself should only be considered if you are fully acquainted with all the necessary (legally enforceable) safety precautions and you are prepared to incorporate them fully in the construction without any compromise. See also Section 1.1.11.

7.1 Energy consumption

Search lights, spots, multiple colour beams, and others, convert the electrical energy applied to them mainly into heat. The efficiency is low: about 3–6 per cent. This means that, say, a search-light of 500 W converts only 15 W into light; the remainder is converted to useless heat. This means that such a searchlight can be compared with a small heater that also gives off some light.

From this, it will be clear that a lighting installation consumes an enormous amount of energy, which it may not always be possible to take from the usual mains outlets. In the United Kingdom, a ring circuit (which is protected by 30 A fuses) can provide a maximum of 7200 W; in many other countries where no ring circuits are available, the maximum power from a mains outlet is usually of the order of 3000–4000 W. Obviously, a large lighting installation must be powered by a 3-phase mains system, and this may not be available in many halls, certainly not the smaller ones. Another aspect that must be borne in mind is that many lighting units draw a high switch-on current, so that they should not all be switched on at the same time to prevent the fuses melting or the cut-outs tripping continuously.

Owing to the high powers, and the consequent large currents, cables and plugs should be of good quality and not be impaired in any way. A bad electrical contact with a high transfer resistance leads to overheating, destruction of the relevant contacts, and is a potential source of fire. For that reason, cables and connectors must be inspected regularly, preferably after

263

each performance.

Cable must be of sufficient cross-sectional diameter (see 5.2.2). In many high-wattage light units, the leads must have special heat-resistant insulation. Cable should not remain wound on a cable reel to prevent it overheating.

7.2 Interference

Most lighting installations unfortunately cause all sorts of crackle, plops and hum that are picked up by the sound installation. Switch-on crackles are caused by the sparking of switches when they are turned on or off. Such interference can be reduced greatly by a shunt capacitor as shown in Figure 7.2.1. Figure 7.2.2 is an oscillogram of the interference caused by a light switch, and Figure 7.2.3 shows how the interference is reduced substantially by the capacitor. Such a capacitor can normally be added to existing switches. Its voltage rating should be at least double the mains voltage.

Interference caused by light dimmers is normally far more serious than switch-on crackles and also much more difficult to get rid of. This is because the current applied by the dimmer to the lighting is no longer sinusoidal, but consists of short pieces of a sine wave and is therefore very rich in harmonics (see Figure 7.2.4). The number and amplitude of the harmonics increases the more the dimmer is turned down, that is, the less light the lamps give. This interference is propagated not only via the mains supply but also by radiation on to electrical wiring. It is very difficult and expensive to eradicate; the costs are really only justifiable in large, sophisticated installations. There are some practical ways to reduce the effect of this interference:

- connect the lighting installation to a different mains supply ring than the sound installation;

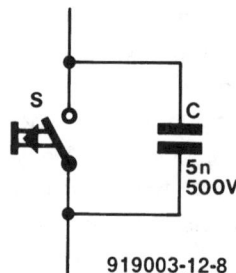

919003-12-8

Figure 7.2.1. Connecting a capacitor in parallel with a light switch greatly reduces interference caused by the switch being turned on or off.

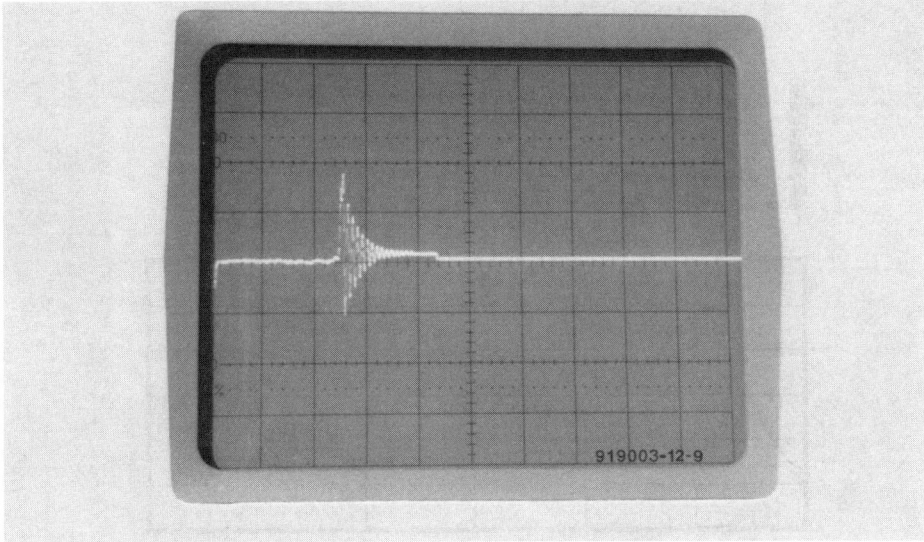

Figure 7.2.2. Oscillogram of interference caused by sparking of switch contacts.

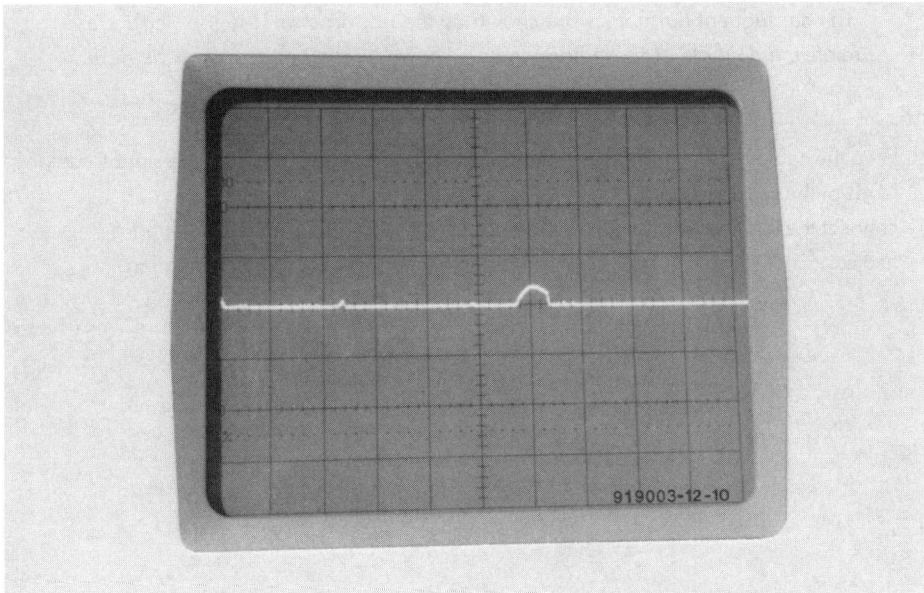

Figure 7.2.3. Oscillogram showing the beneficial effect of a capacitor
on the interference caused by sparking of switch contacts.

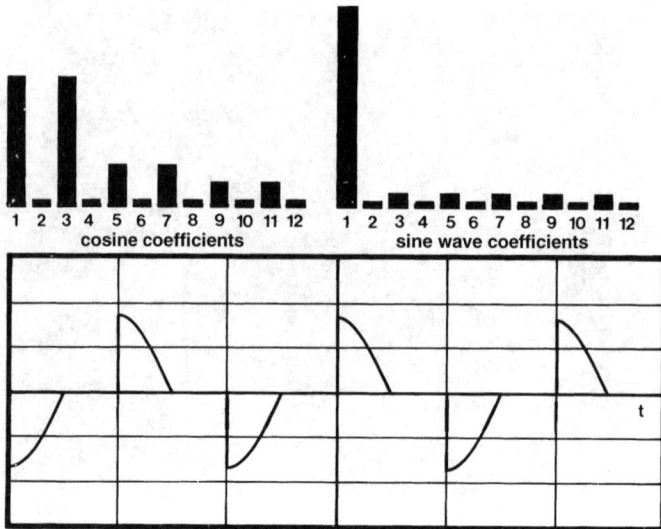

Figure 7.2.4. Waveform of the current supplied by a dimmer to the lamp.
The consequent harmonics are shown by the bar graph at the top (left); the
relative amplitude of these harmonics is shown by the bar graph at the right.

- keep the cables to the lighting installation away as far as possible from the signal-carrying leads of the sound installation;
- connect the lighting installation to the mains supply via a suitable mains filter;
- fit a capacitor (as in Figure 7.2.1) across all switches in the lighting installation.

8. Practice area

Not all rooms or halls are suitable for practice sessions: some specific matters must be borne in mind. Often, practice space can be hired in partially empty public buildings. Bear in minds that every time there is a practice session, percussion, amplifiers, loudspeakers, and much more will have to be transported, so accessibility and distance are factors to be considered. Also, a practice somewhere on the third, fourth or fifth floor is not a very attractive proposition, unless the building has suitable lifts. The sound volume is, of course, also a factor, but this is more for the consideration of the landlord.

8.1 Sound insulation

Live music is loud, whether it concerns a classical orchestra or a pop group. When it is considered that the pianissimo parts must be clearly audible, it will be obvious that with the large dynamic range of live music the fortissimo passages will be very loud, indeed.

To prevent the sound becoming a nuisance to neighbouring residents, it is essential that as little sound is audible outside as possible. Sound absorption is also desirable for musical reasons. The practice space should not sound hollow (reverb) since this results in an undefined sound image which makes effective practising impossible. So, reflections of sound waves must be minimized. This means that suitable sound absorbers must be used on the floor, against the walls and to the ceiling. There are modular sound absorbers available that consist of small, square boxes of various depths in which cavities are formed by suitable dividers over which a layer of mineral wool is fixed together with a perforated faceboard. By altering the percentage perforations, the thickness and density of the wool, and the depth of the cavities, the absorption characteristics can be suitably modified. These absorbers are particularly useful in small to medium sized rooms or halls.

If money is scarce, egg boxes and old pieces of carpet can be used to form good sound insulation. The egg boxes form hollow spaces (see Figure 8.1.1) in which sound waves are trapped, attenuate one another and move the carton.

In pieces of carpet, the sound energy is converted into kinetic energy of the carpet fibres. Since no hollow spaces are created, the attenuation at low and middle frequencies is not as good as provided by egg boxes.

Not all walls, floor and ceiling should be covered with absorbing material, however, since the room should then sound lifeless. Therefore, one wall, or the ceiling, should not be covered with absorbing material.

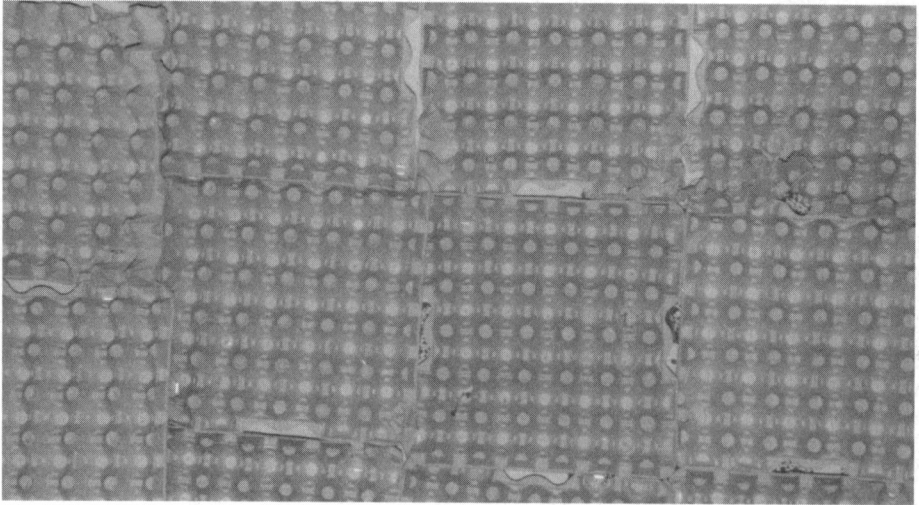

*Figure 8.1.1. Egg boxes form good (and cheap) sound absorber,
but not at low frequencies.*

The use of egg boxes and pieces of carpet is effective only at middle and high frequencies: it is much more difficult to attenuate low frequencies.

The effectiveness of various materials used for sound insulation is shown in the frequency vs attenuation characteristics in Figure 8.1.2.

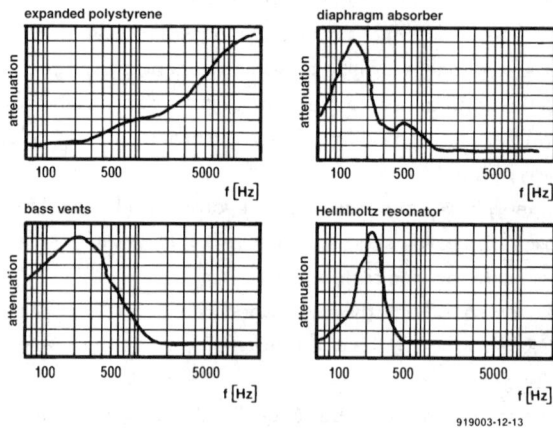

919003-12-13

Figure 8.1.2. Comparison of various sound absorbing materials.

8.2 Mains power

The power provision, particularly in older buildings (where practice areas are often situated) can create problems. Two or three different ring circuits, each with a number of socket outlets, would be ideal. This would make it possible for the group or band to be split into a couple of sections to lessen the load on each socket outlet and also to reduce any interference.

If there are not enough socket outlets available, a two-gang or four-gang (fused) extension socket as shown in Figure 8.2.1 should be used, but care should be taken not to exceed the loading of the outlet into which this is plugged.

919003-12-14

Figure 8.2.1. Fused four-gang extension socket to BS1363/A.

8.3 Recording in the practice area

Reasonably good-quality tape recordings may be made with simple means in a practice area that is acoustically well attenuated. This may be done in a number of ways: the entire performance can be recorded at once, or groups of instruments may be recorded in turn. The best results are obtained with a recording of the entire performance in one session.

8.3.1 Tape recording

In a tape recording, the tape noise should always be taken into account. It arises because of the granular nature of the tape coating material and the magnetic domains within it. There are also bias noise and contact noise (due to the surface nature of the tape) to be considered. Good-quality tape and skilled recording gives a signal-to-noise ratio in reel-to-reel recorders of about

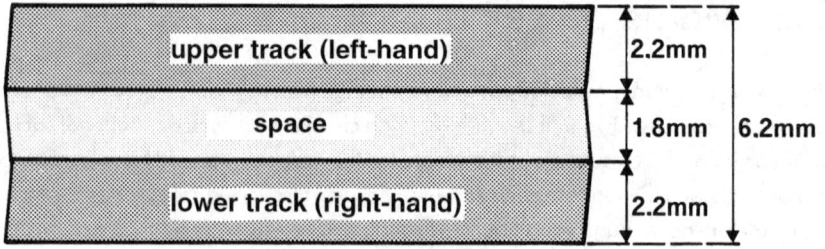

Figure 8.3.1. Track layout in a two-track tape recorder.

80 dB. Using similar tape in a cassette recorder, the S/N ratio is unlikely to be better than about 60 dB. Noise reduction techniques may improve these figure to some extent.

Only tape recorders that provide high tape speeds (at least 19 cm/s or 7.5 in/s) should be used: the usual tape speed of a good reel-to-reel recorder is 15 in/s (38 cm/s). Note that owing to the predominance of the USA in the recording field, metric dimensions are very seldom used in specifying tape widths or tape speeds.

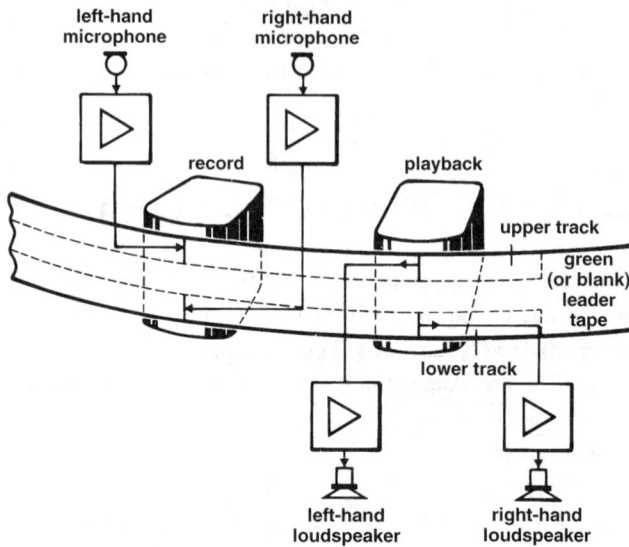

919003-12-16

Figure 8.3.2. Basic layout of a two-track reel-to-reel tape recorder

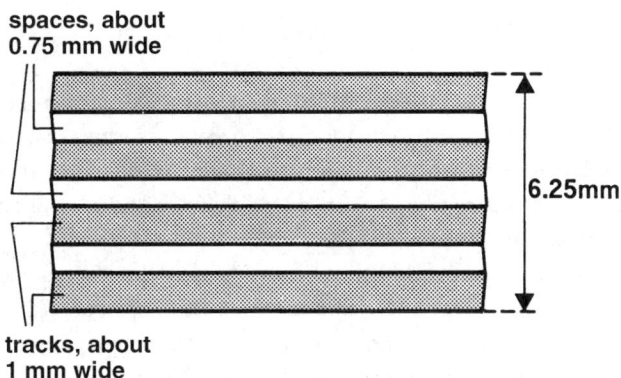

spaces, about
0.75 mm wide

6.25mm

tracks, about
1 mm wide

919003-12-17

Figure 8.3.3. Track layout in a four-track reel-to-reel tape recorder.

The magnetic tape is a continuous strip of durable plastic which is given a uniform coating (about 90 μm thick)of magnetizable material, such as ferric oxide or chromium oxide. Good tapes and tape recorders have a dynamic range of up to 120 dB; cassette tape and recorders up to about 70 dB.

When stereo tape recording became popular (mono tape recording started in the 1940s), the two-track system was the most frequently encountered. In this system, two tracks, each 2.2 mm wide, separated by a 'dead band' of 1.8 mm, are magnetized in the same direction (see Figure 8.3.1). The upper track is used for the left-hand channel, and the lower one for the right-hand channel (see Figure 8.3.2). Because of its good S/N ratio and the immunity to dirt on the tape, this system is still widely used in recording studios .

To meet the requirements for a stereo system in which the tape need not be wound back repeatedly, the four-track tape recorder for domestic hi-fi applications was developed. It has the advantage of offering twice the playback time for the same length of tape, and the tape need not be wound back. Since four tracks need three dead bands, they are only about 1 mm wide (see Figure 8.3.3).

Before each and every recording, the recording level must be established with the aid of the VU meter. To obtain as large a S/N ratio as possible, the tape drive must give maximum deflection of the VU meter, but there should be no overdrive even during the loudest passages. Some recorders are equipped with automatic recording control (ARC), which may be used if mediocre quality recordings are acceptable. Such an ARC depends on a kind of dynamic range compressor, which compresses the recorded material and which is, therefore, not suitable for first class recordings.

8.3.2 Direct recording with spaced microphones

The simplest way of making highly usable recordings is with spaced microphones (A-B system) already described in Section 6.1.2. A standard stereo reel-to-reel tape recorder (operating with a tape speed of 15 in/s) and two good-quality recording microphones are needed. The quality of the recording depends to a large extent on the placing of the microphones and the instruments in the practice area.

A more or less circular arrangement as shown in Figure 8.3.4 gives good results in practice sessions. The arrangement ensures that each musician can hear all the other instruments clearly. To make a good tape recording, the arrangement must be changed, however. It is then necessary that each instrument is picked up by the microphone(s) at the same level. It has been found that a semicircular or straight-line arrangement as shown in Figure 8.3.5 is then better. During the recording, it is important to ensure that the microphones are not placed at the crest of a standing wave (see Section 1.3). When this happens, the VU meter deflects excessively at certain – normally low – frequencies. The relevant microphone should be repositioned.

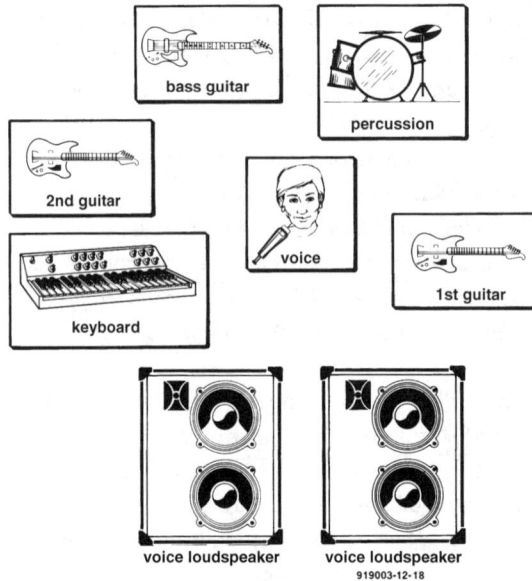

Figure 8.3.4. It has been found that for practice sessions this circular arrangement works well for all musicians.

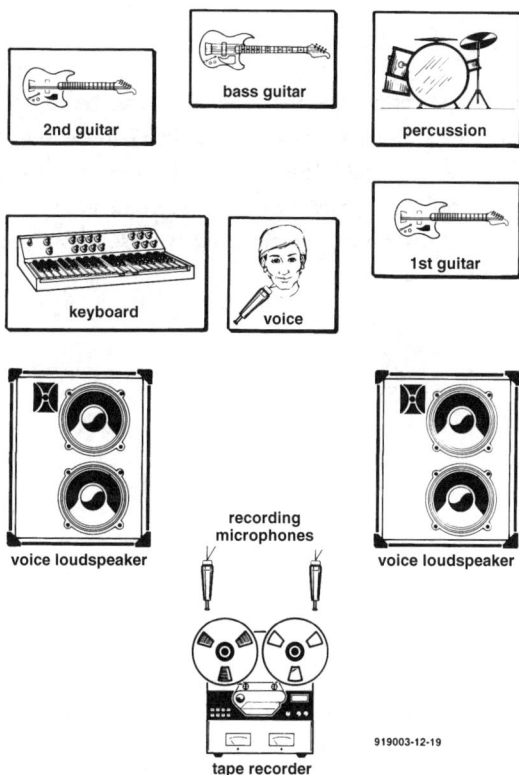

Figure 8.3.5. During tape recording it is important that both microphones can pick up all the instruments. For this, a semicircular or straight-line arrangement of the musicians has proved to be best.

8.3.3 Multi-play recording

Some tape recorders allow multi-play recordings to be made. This enables different instruments to be recorded one after another, albeit monophonic. It is done as follows: a section of the instruments, say, the percussion, is recorded on one track. This recording is then copied to another track, while at the same time another section of instruments is recorded (remix). In theory, this process could be repeated ad infinitum: every time a copy is made, another instrument or instruments are added, while the result is monitored by headphones (to ensure synchrony). In practice, more than five copies are not recommended to prevent the noise level becoming unacceptably high.

8.3.4 Multi-track recording

Although there is still a place for single-channel tape recorders, the bulk of recordings are now made on multi-track machines: 8-track, 16-track or even 32-track. This is of particular interest in the pop music field, where individual instrumentalists or singers in a group normally perform separately in sound-proof rooms. Each of these is recorded on a separate track from which the stereo master is then made by mixing and blending of the separate tracks. All sorts of effect can be added at this stage.

9. Faultfinding and small repairs

Where much technical equipment is used, faults may occur and apparatus may fail, and the incidence of these is in direct proportion to how much heavy use it gets – in a pop group normally quite a lot. Fortunately, most damage is not of the serious kind for which the assistance of a specialist repairman has to be sought: many faults are trifles which can usually be remedied readily on the spot.

9.1 Wonky contacts in cables and connectors

Cables and leads used by music groups, whether for the guitar microphone or for the power supply, are subject to heavy wear and tear. This is a good reason for using robust connectors, such as 6.3 mm metal audio plugs and sockets: plastic ones do not last very long, particularly if someone stands on them. Nevertheless, many faults are caused by a break in a cable or a cable pulled out of a plug or socket. In case of a break, the exact spot where this occurs must be found by 'bending and feeling' of the cable (see also Section 5.2). If the cable has been pulled out of a plug or socket, it has to be reconnected.

9.2 Faulty equipment

Nowadays, an equipment will seldom fail because of an internal defect. Mostly, the cause will be a cable pulled out of a plug or socket or a faulty mains cable. When, for instance, an amplifier fails, the first thing to do is to check all the connections thoroughly. Only when there is no doubt that all connections are sound and there is no question of an operating error, is it likely that the amplifier itself is faulty. Normally, this is caused by a blown fuse or a faulty valve. Both these are easily replaced in situ. Mechanical operating controls such as switches and potentiometers, as well as indicator lights, can usually be replaced fairly easily. More serious faults should be left to specialized technicians.

9.2.1 Blown fuses

When an equipment does not show any signs of life after it has been switched on, or indicator lamps and LEDs do not light, it may be that the mains fuse at the back of the unit or that in the mains plug (in United Kingdom only) has blown. The fuse at the back of the unit serves two

275

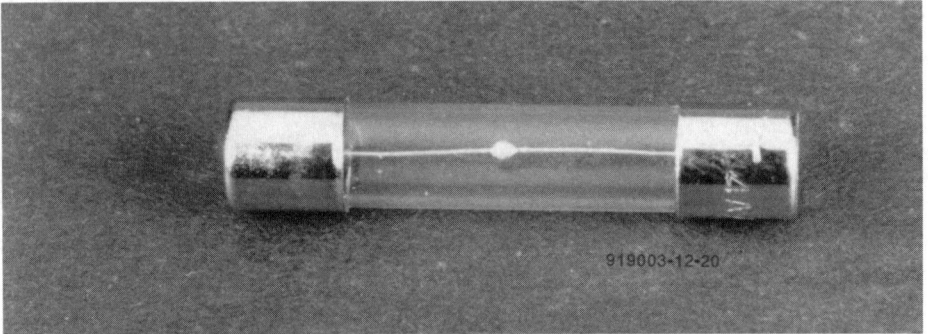

Figure 9.2.1. Because of its low cost, a glass fuse as shown is the most frequently used type in electrical and electronic equipment

functions:

- when a fault occurs, it must prevent a fire (by interrupting the supply line);
- it protects the equipment against more serious damage which may result from the original fault.

The fuses used in electronic equipment are invariably of the glass type as shown in Figure 9.2.1. When a certain current level (the rating of the fuse) is exceeded, a thin wire in the fuse becomes so hot that it melts and so interrupts the supply line in which it is located. The construction of these fuses is universal: a small glass tube that contains a thin wire which is welded to the two metal caps that close the tube. In some fuses, the tube is filled with fine sand, which in case of the wire melting prevents any sparks. The data of the fuse are indelibly shown on one of the endcaps or on the glass tube itself. These data normally are:

- the maximum voltage that may be applied across it;
- the rating in amperes or milliamperes at which the fuse blows;
- the speed at which the fuse acts.

The first two are shown in the form of a number, but for the speed an abbreviation is used:

FF = very fast
F = fast
M = slowish
T = slow
TT = very slow

Very fast fuses blow immediately their rating is exceeded, whereas very slow fuses blow only after the excessive current has flown for some time. At ten times the rated current, a very fast fuse blows 500 times faster than a very slow one.

Basically, a fuse must blow rapidly, but slow fuses are needed in supply leads to transformers or reservoir capacitors where rapid action is not required. If, in such cases, a slow fuse would be replaced by a fast one, the fuse would blow every time the supply is switched on. If, on the other hand, a fast fuse is replaced by a slow one, the relevant equipment may no longer be adequately protected.

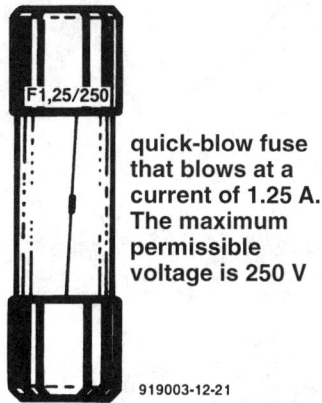

quick-blow fuse that blows at a current of 1.25 A. The maximum permissible voltage is 250 V

919003-12-21

Figure 9.2.2. Example of the indelible print on a glass fuse.

An example of the printing on the fuse is shown in Figure 9.2.2. The dimensions of European fuses are standardized at 5×20 mm; Japanese and American types are slightly larger at 6.3×32 mm.

It often happens that a fuse blows although the relevant equipment is not faulty. This may be caused by voltage spikes on the mains supply or by a faulty valve. This is the reason why a sound technician should have not only a number of spare valves in his toolkit, but also a good supply of spare fuses. Obviously, we only need spare fuses for those in actual use: so, a note should be made of the ratings of all the fuses used in an installation. By the way, before a fuse is replaced, pull the mains plug from the socket – it's safer! Clearly, the blown fuse must be replaced by an identical type.

If the equipment is really faulty, replacing a fuse does not help, of course. So, if a fuse that has just been replaced blows immediately, it does not need replacing again, because this points to a serious fault for which the equipment has to go to a repair workshop.

Figure 9.2.3. The mains fuse is normally located at the back of an equipment: it is accessible after a screwcap has been removed. In the guitar amplifier at the top, the fuse holder is fitted to the chassis. After the cap has been removed, the equipment can optionally be set for operation at different mains voltages. The illustration at the left is the mains entry of an active loudspeaker. Here, the fuse is located to the right of the mains entry. The cap can be unscrewed with a small coin. The connectors at the bottom are intended for keyboard leads and have their own internal fuse.

9.2.2 Faulty valves

In contrast to transistors and integrated circuits (ICs), valves have a limited life. Depending on how and where the valve is used, its life may be months or years; the actual life depends on a number of factors:

- the quality of the valve;
- the number of times the equipment in which it is used is switched on and off;
- the total time the equipment in which it is used is operational;
- the design of the amplifier.

The first of these factors has already been discussed in Section 3.3.5. When the equipment is regularly switched on and off, the heater of the valve undergoes a lot of wear and tear. A stand-by switch would be of great benefit in such an equipment. An equipment without a standby switch is better left on during intervals and interruptions since this would be of benefit to the valves it uses. When an equipment is on standby or is operating, the glowing of the heater of the valve shows that it is in working order. When a heater is faulty, it does not glow.

As mentioned in an earlier chapter (3.3.5), the length of the periods the equipment is operated has, of course, an effect on the life of the valves. Another aspect that has an effect on the life of the valves is the design of the amplifier. Valves are worked much harder in a Class A amplifier than in a Class B amplifier. The power dissipation in valves in a Class A amplifier when there is no input signal is very high: some amplifiers get very hot in these conditions. In spite of earlier statements, it is advisable to switch off such amplifiers during intervals in the performance and accept the consequences on the heaters.

Old, worn out valves in the preamplifier cause a reduction in the rated output power and the amplifier's capability of being overdriven. When output valves are reaching the end of their lives can be heard. Most amplifiers use push-pull output stages in which two or four valves work together. Obviously, the parameters of these valves must be identical or nearly so. When one of the valves becomes faulty, the operating point of the other(s) is shifted, whereupon the sound of the output changes: it becomes grainy or coarse. To prevent unbalanced operation of the push-pull stages, all valves in it must be replaced together even if only one is faulty.

A valve with a faulty internal electrode system is easily located by softly tapping the valves as described in Section 3.3.5.

9.2.3 Faults in the equipment

Statistically, the following four (in order of 'merit') cause faults in an equipment more often than others:

- power supply;
- transistors and integrated circuits in the power sections;
- electromechanical parts, such as switches and potentiometers;
- preamplifier stages.

The power supply is far and away the most frequent culprit. Typical of a faulty power supply is that 'nothing works any more': the equipment does not react to the operation of any controls, no lamps or LEDs light. In fact, they are the same kind of symptoms as when a fuse has blown.

When a musician's amplifier only produces a soft hiss, it usually indicates a faulty preamplifier.

If you have no or not much experience of electronic equipment, and an amplifier or mixer is suspected of having an internal fault, it is best to take it to a competent service department. Unfortunately, in the world of music, many of these workshops have staff that, to put it mildly, do not know a capacitor from a resistor. Therefore, before leaving an equipment in such a department, discuss the fault with one of the service technicians: this will give you a good idea of his/her competence. Also, look around in the workshop to get an idea of the standards of workmanship: in a well-run one this will not be refused.

When a good workshop has been found with competent technicians, a detailed description of the circumstances in which the equipment became faulty and the symptoms of the fault can often save repair time and thus money. In any case, before giving instructions to carry out the service, get an estimate. Finally, when collecting the repaired equipment, inquire after the fault and what the possible causes for it were: this knowledge may prevent a repeat.

9.3 Faulty instruments

The electronics in an electric lead guitar or bass guitar is generally not very complicated, so that little can go wrong there. The most usual faults are noisy potentiometers, loose audio sockets and bad contacts. A good, non-microphonic pick-up cannot really become faulty unless it is badly mishandled since the turns are carefully fixed in place with resin. Pick-ups with loose windings may suffer from a break in the wire.

Faulty potentiometers are easily recognized: they make an awful noise when they are turned. When a potentiometer is soldered directly to a printed-circuit board, it may happen that a solder joint fails, and this also causes crackles. The remedy is, of course, to resolder the joint. If the solder joints are all sound, the noise may be caused by dirt and dust on the track(s) of the potentiometer: this may be removed with a proprietary contact spray. This may have to be repeated regularly to prevent a recurrence of the fault. It is advisable to cover equipment with a hose or just plain cloth to protect it from dust.

Because of the heavy work a wah-wah pedal does, the potentiometers in it often become faulty. The moving parts of the pedal should be inspected regularly and given a thin coating of acid-free grease.

Dirty switches and sockets may also be cleansed with a good contact spray. When the spray has been applied, operate the switch a couple of times. Clearly, the mains supply to the equipment should be switched off before this work is undertaken. After the spray treatment, leave the equipment for about 15 minutes before switching on the mains again: this gives the spray remnants time to ev evaporate.

Often used knobs and potentiometers may need to be replaced after a certain period of time. It is advisable, if at all possible, to replace them with the best quality and sturdiest types available.

9.4 Faulty microphones and loudspeakers

It is fortunately rare for microphones and loudspeakers to become defective, which is just as well, because, owing to the specialized measuring instruments and tools needed for inspection and repair, there is not much that can be done outside a suitable workshop. It is, of course, always possible that the fault is simply a bad contact in a microphone cable or plug, and this can be remedied in situ.

Often, it may not be evident whether the output amplifier is faulty or the loudspeaker. This can be ascertained readily by connecting another (identical) loudspeaker to the amplifier.

Faultfinding tables

1 Valve amplifiers

symptom	possible cause	remedy
no sound; indicator lamps and LEDs do not light	mains lead faulty or not plugged into mains socket; power supply faulty	repair or connect as appropriate
only one channel usable	valve(s) in preamplifier faulty or signal path of channel interrupted	replace valves or repair signal path interruption
audio signal weak; no or little overdrive possible	preamplifier valves worn; faulty power supply	replace valves or repair power supply

symptom	possible cause	remedy
normal audio signal level but with crackles	faulty valve; bad solder joint(s)	check valves by gently tapping; resolder suspect joint(s)
normal audio signal level but with hum	faulty power supply	take equipment to repair workshop
equipment howls when a certain signal level is exceeded	positive feedback caused by defective valve; insufficient screening of valves; feedback circuit faulty	fit screening cans on valves; take equipment to repair workshop

1 Solid-state amplifiers

symptom	possible cause	remedy
amplifier produces no sound; indicator lamps and LEDs do not light	mains lead faulty or not connected	repair or connect as appropriate
only one channel usable	preamplifier faulty or signal path of channel interrupted	take equipment to repair workshop
audio signal weak; no or little overdrive possible	faulty preamplifier or output amplifier	take equipment to repair workshop
normal audio signal level but with unwanted distortion	faulty output amplifier or power supply	take equipment to repair workshop
normal audio signal level but with crackles	poor solder joint(s)	resolder joint(s); take equipment to repair workshop
normal audio signal level but with hum	defective power supply	take equipment to repair workshop
equipment howls when a certain signal level is exceeded	acoustic feedback; fault in feedback circuit	take equipment to repair workshop

10. Studio recording

A recordings made in a practice area with simple equipment is often not good enough, particularly when it is intended to be sent to an agent or record company. Most of us are thoroughly spoilt by the quality of radio, television and compact discs. In his mind, a layman cannot divorce tape noise and limited frequency range from the musical skills of the performers. To the layman, sound quality is the only factor that makes it worth listening to the music, and for him/her the quality of a CD reproduction is the standard.

The foregoing is the reason that most bands go to a studio to make a recording. Today, there are untold numbers of studios in most towns and cities. Because of the rivalry between them, rentals, including the services of a sound technician, are very modest.

Unfortunately, there are willing laymen who, since they are surrounded by modern and expensive equipment, think that they are skilled sound technicians. A high-quality mixer and extensive peripheral equipment are no substitute, however, for skill and, therefore, no guarantee for a successful recording. A face-to-face discussion and a short inspection before hiring a studio can prevent much disappointment and save money.

Most studios charge a fixed amount per hour plus the cost of any material used, such as tapes. It is, therefore, important that the recording is thoroughly prepared and rehearsed. It is not enough for all musicians to rehearse together: each and every individual performer must know his part inside out and be able to play it absolute faultlessly. Such preparation can save much money, time and effort in the studio.

10.1 Setup of a recording studio

Figure 10.1 shows the layout of two small recording studios. Such a studio must consist of at least a recording room and a control room (where the mixer is situated).

The recording and control rooms in the studio in Figure 10.1.1a are completely isolated: communication between the two is via an intercom (microphone and headphones). Because of the isolation, the recording room sounds dry. There is a separate, sound-proof room in which, for instance, percussion instruments may be recorded via separate microphones.

The mixer, various effects units and a tape recorder are installed in the control room. Since this room is relatively large, it may also be used for direct recording of keyboard instruments, such as synthesizers. In view of the small distance, the use of DI boxes is not necessary, but not impossible.

The studio in Figure 10.1.1b has a glass partition between the recording and control rooms

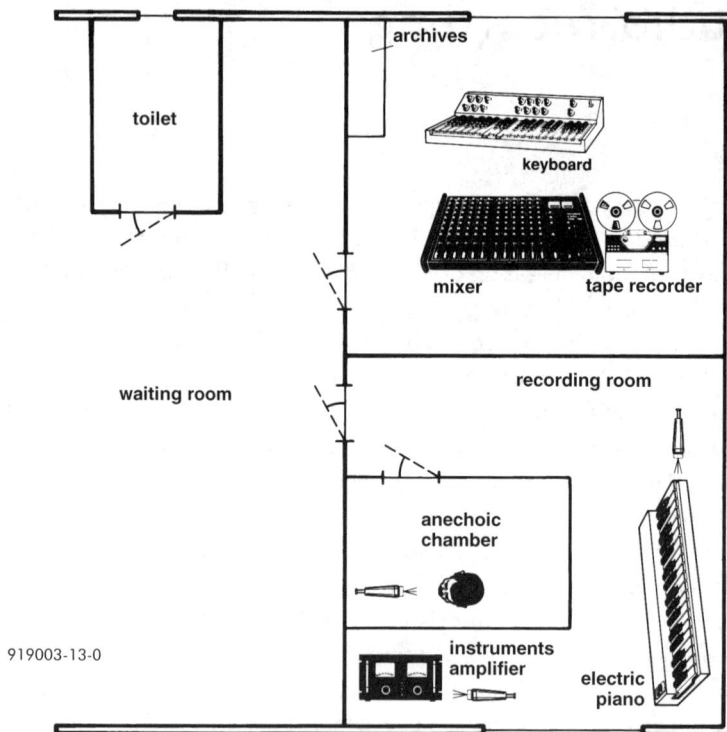

Figure 10.1.1a. Layout of recording studio with isolated
recording room and control room.

to make it possible for visual contact to be maintained. There is also a large reverb room in which, say, percussion instruments that are recorded with spaced microphones can be given natural reverberation during the recording.

10.1.1 The recording area

The most important property of a recording area is good sound insulation to reduce unwanted sounds entering the area. Sound travels via many paths, but air gaps around windows, doors, and through ventilation ducts are particularly difficult to insulate. Extraneous sounds would colour the sound, which would degrade the recorded music. Figure 10.1.1 shows an example of a well-insulated room: the whole floor is covered by carpeting and the walls are covered with special sound absorbing panels: all this guarantees an acoustically dry area. Note in the

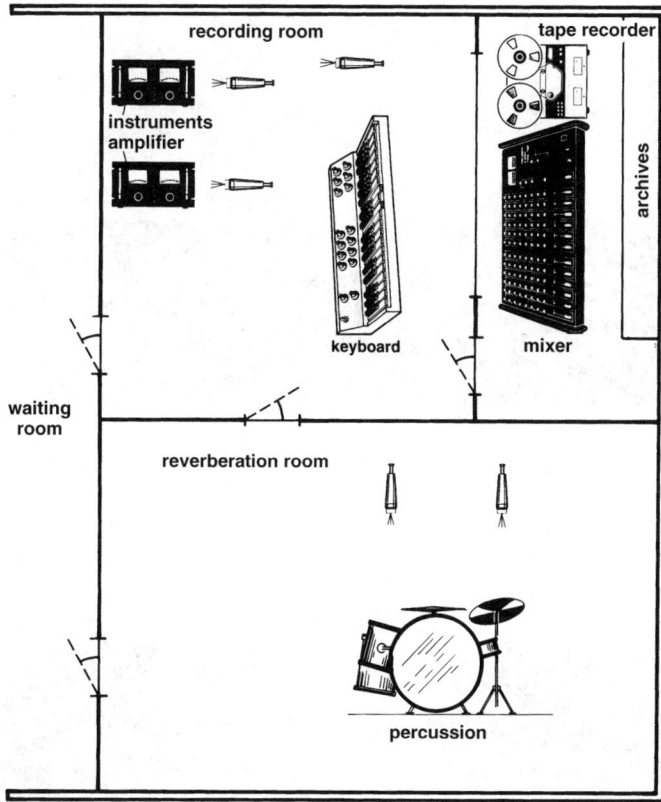

Figure 10.1.1b. Layout of recording studio with glass partition between recording room and control room.

illustration that there is a space between the panels and the wall, which is effective in damping low frequencies.

The recording studio should have a large number of power points around the rooms. A well-equipped studio also has available a number of standard instruments and equipment, such as an electric piano, synthesizer, drum computers, and others (see Figure 10.1.3).

The connections between recording and control rooms are balanced. There is, or should be, a special connecting box for microphones and headphones (see Figure 10.1.4). The box in the photograph has four stereo phono chassis sockets for microphones, which provide balanced operation. Although unusual, there is nothing against the use of phono sockets instead of XLR sockets.

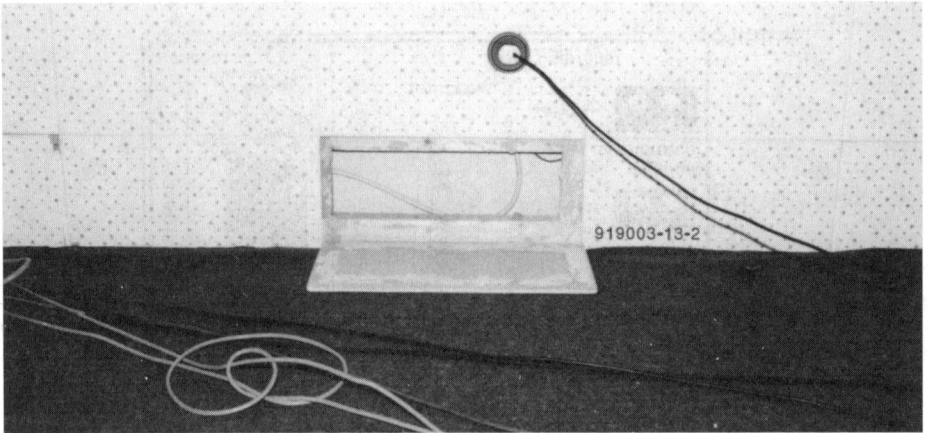

Figure 10.1.2. Typical sound insulation in a recording area: carpeting on the floor and sound-absorbing panels on the walls. One such panel has been removed to show the space between it and the wall (in which cables may be stowed).

Figure 10.1.3. In many studios, a drum computer is standard equipment nowadays. It may be used for synchronization during a recording session, but is not a substitute for acoustic percussion.

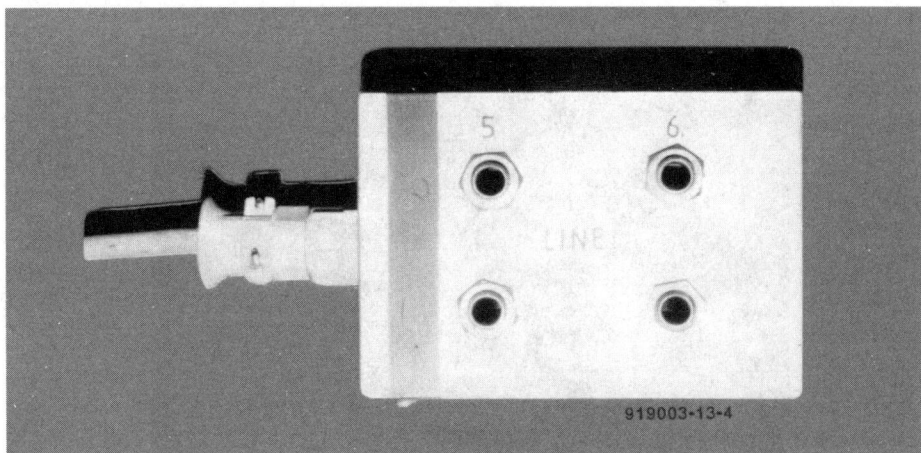

Figure 10.1.4. Connecting box for four microphones; unusually, stereo phono chassis sockets instead of the more usual XLR sockets are used.

10.1.2 The control room

All the operating equipment of the recording studio is kept in the control room. This consists of at least a mixer, a multi-track tape recorder, various effects units, a master tape recorder and possibly a MIDI computer for controlling modern keyboard instruments.

The multi-track tape recorder is the core of the installation. The number of tracks of this machine determines how many instruments can be mixed separately or in groups. In fact, therefore, this

Tape speed tolerance	±0.2%
Pitch control	±15%
Tape width	1 in
Tape speed	15 in/sec
Frequency range	40–22,000 Hz, ±3 dB at 0 VU
	40–22,000 Hz, ±2 dB at –10 VU
Distortion factor	0.8% at 0 V VU and 1 kHz
	3% at 13 dB above VU
Signal-to-noise ratio	69 dB; 107 dB with dbx

Figure 10.1.5. Brief technical parameters of a 16-track reel-to-reel tape recorder. The specification is typical of a machine used by tape recording producers.

Figure 10.1.6. Typical 16-track tape recorder (top) and associated input channel control and VU meters (bottom)

recorder determines to a large extent what the possibilities in the studio are. It is, of course, a very special recorder that uses 1/2 in (8 tracks), 1in (16 tracks) or 2 in (32 tracks) wide tape and enables 8, 16, or 32 tracks to be recorded separately of each other. The tape speed is 15 in/s. The brief technical details in Figure 10.1.5 refer to a 16-track recorder and show that even this kind of professional machine causes some losses. A 16-track recorder and the associated input controls with VU meters is illustrated in Figure 10.1.6.

Figure 10.1.7. Studio cum public-address mixer from Dynacord.

Figure 10.1.8. Large studio mixer from Dynacord.

Obviously, a tape recorder without a mixer is all but useless; it is the combination of the two units that enable a music performance to be processed and produced successfully. Studio mixers do not differ much from those in public-address systems, but in a studio more facilities are needed to form individual channels into groups—see Figures 10.1.7 and 10.1.8.

Finally, there are the effects units which, just as in a public-address system, are connected in the effects path of the mixer. Since in a studio sound is damped appreciably, reverb and echo are produced artificially and added to the recording. In that way, the recording can be given a spatial impression within wide limits.

The instruments are recorded on their own or in small groups. Each instrument or group of instruments is allocated one or two tracks by the mixer. A recording with two microphones on two discrete tracks may be advantageous in the case of instruments that are highly directive. If, for instance, an electric guitar is recorded from the guitar loudspeaker, the sound will be unusually sharp and not resemble that heard in the practice room. On the other hand, a guitar that is recorded wholly indirectly will lose presence and clarity. For that reason, it is better to use two microphones. One microphone is pointed directly at the source of sound from a short distance, while the other is placed further away in the room. The combined signal they pick up sounds more natural than would be obtained if the sound were recorded from just one position. This is true with other instruments as well, of course.

The percussion section always needs at least two tracks. When discrete microphones are used, each part of the percussion may be given its own track, or use can be made of a sub-mixer, or the percussion is combined into a sub-group at the main mixer. This last method is the least flexible for the final recording because there are then already several instruments on one track.

When all instruments have been recorded, the tracks are combined via the mixer; at this stage, effects can be added. The editing normally takes as much time as the recordings. When this is done, the result is copied on to a master tape, which is run on a two-track stereo reel-to-reel recorder at a speed of 15 in/s.

10.2 Noise suppression

Owing to the multi-track system of the first recordings, noise is added to the music signal. It is therefore advisable to switch on the noise reduction system of the tape recorder. The same is true when the master tape is being copied on to a cassette or other tape. There are various noise reduction systems in use.

* *Dynamic noise limiter (DNL)*. This is based on simple equalization. Since the human ear is not as sensitive to some frequencies as to others, the frequency response of a suitable filter is made to provide progressively more high-frequency cut as the sig-

nal level falls to mimic the response of human hearing. It should be noted that DNL is active only during playback, since the tape recording itself has not been modified. The recording can therefore be played back with or without the DNL switched on.

- *dbx*, of which there are two versions: Type I for high bandwidth (professional) systems and Type II for limited bandwidth (cassette tape) systems. Although the two are not compatible, they both use 2:1 compressors and 1:2 expanders (see Chapter 5). During recording, the music signal is compressed in a ratio of 2:1, which alters the frequency range, while during playback the signal is expanded (decompressed) by a ratio 1:2 and the original frequency range restored.
- *Dolby A.* Basically, this resembles dbx, but the signal compression is carried out frequency-dependent in four different frequency bands: <80 Hz; 80 Hz to 3 kHz; 3–9 kHz; 9 kHz. Both Dolby and dbx must be used only once: when a recording that has been made with Dolby or dbx switched on is being copied, the Dolby or dbx system must be switched off. If this is not done, the noise would be amplified and the frequency response irretrievably curved. The operation of the Dolby noise reduction system is represented in simplified form in Figure 10.2.1.

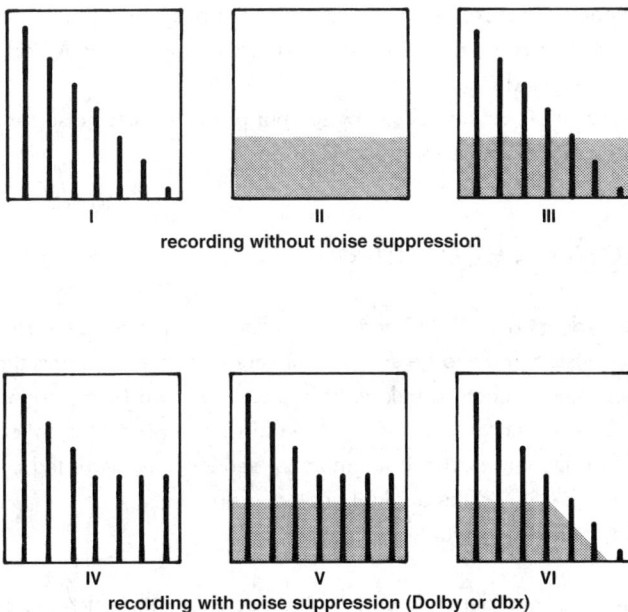

recording without noise suppression

recording with noise suppression (Dolby or dbx)

Figure 10.2.1. Simplified representation of the operation of
the Dolby noise reduction system.

In I, the audio signal starts loudly and then tapers off.

In II, the tape noise is represented by a grey bar.

III shows how during a recording, the soft music passages are submerged in the tape noise, which is audible even during intervals. The noise is masked by the louder passages only but, since its character differs from that of the music, it may remain audible.

IV. During the recording with the Dolby system switched on, soft passages are amplified. The amplification may be frequency-dependent.

V shows what is present on the tape: owing to the amplification, the soft passages have risen above the tape noise floor. The frequency response no longer resembles that of the original signal.

VI. During playback, all passages that were amplified during the recording are attenuated proportionally; at the same time, the tape noise is also attenuated. The result is that the soft passages are still well clear of the noise floor.

- *Dolby B* has become the standard noise reduction system for pre-recorded cassette tapes and is therefore only of academic interest here.
- *Dolby C* is based on Dolby A and B, but increases the amount of noise reduction to 20 dB.
- *Dolby SR* (Spectral Recording) is a professional noise reduction system based on Dolby A, B and C (a combination of fixed band companders as A and sliding band companders as in B and C).
- *Dolby S* is a domestic version of Dolby SR, but produces less noise reduction at low frequencies than SR, but more than B or C.

10.3 Digital recording

Today, master recordings are all digitized (also called quantized). The basic technique is for the analogue music signals to be sampled at very brief intervals, and then for the instantaneous peak value of each sample to be represented by a binary coded number sequence consisting of a train of 0s and 1s (called bits). The process is called Pulse Code Modulation or PCM. During playback, the number sequence is converted by a digital-to-analogue converter (DAC) into the original analogue signal.

The process offers two significant advantages.

- The signal-to-noise ratio depends only on the accuracy with which the signal is sampled. The noise level drops the higher the sampling rate. When a 16-bit resolution (meaning a sampling rate of $2^{16} = 65\,536$ samples/sec) is used, a signal-to-noise ratio of up to 100 dB is possible, while the dynamic range is 90 dB.

- When the master tape is copied, the sound quality remains at the same level, since the signals are regenerated only during playback. Each copy therefore is of the same quality as the original recording. This makes noise reduction systems superfluous.

There are drawbacks, however. One is that more electronic equipment is required for digital recording than for analogue recording. Another is that the sampling rate must be carried out at a frequency at least twice as high as the highest audio frequency to be reproduced. This means that if audio signals up to 20 kHz are to be recorded, the sampling rate must be at least 40 kHz and the tape recorder must be capable of processing such high frequencies.

All tape has, to a degree depending on its quality, little blemishes and imperfections (pollywogs) that result in so-called drop-outs (brief interruptions in the replayed signal). In analogue recordings, these imperfections are normally only disturbing when the tape quality is poor. Digital recordings, however, are very sensitive to them, which makes extensive error correction necessary.

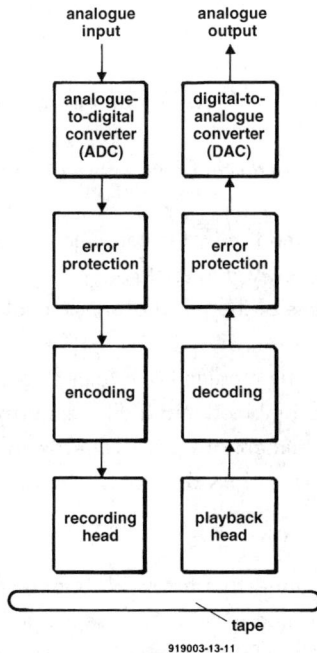

```
   analogue          analogue
    input             output
      |                 ^
      v                 |
 ┌──────────┐     ┌──────────┐
 │analogue- │     │digital-to-│
 │to-digital│     │analogue  │
 │converter │     │converter │
 │  (ADC)   │     │  (DAC)   │
 └──────────┘     └──────────┘
      |                 ^
      v                 |
 ┌──────────┐     ┌──────────┐
 │  error   │     │  error   │
 │protection│     │protection│
 └──────────┘     └──────────┘
      |                 ^
      v                 |
 ┌──────────┐     ┌──────────┐
 │ encoding │     │ decoding │
 └──────────┘     └──────────┘
      |                 ^
      v                 |
 ┌──────────┐     ┌──────────┐
 │recording │     │ playback │
 │   head   │     │   head   │
 └──────────┘     └──────────┘
```

tape

919003-13-11

Figure 10.3.1. Principle of digital recording system. Most of the electronics is contained in analogue-to-digital converters (ADC), digital-to-analogue converters (DAC), and error correction circuits.

Editing a digital tape is a much more difficult process than that of an analogue tape, and, in fact, requires electronic assistance. For this purpose, a range of computer-aided editing systems has been developed in one of which the entire signal is transferred to the hard disk of a computer (which needs a capacity of at least 1 Gbyte), whereupon the editing is carried out with an appropriate computer program.

10.4 MIDI

Not only can home PCs (personal computers) record audio from microphones or line inputs and store them on hard, floppy or removable disk, but most of them have a Musical Instrument Digital Interface – MIDI. This interface enables a computer to control, via a suitable program (software), electronic instruments such as drumsets, synthesizers, and electric pianos and organs.

Information about the piece of music is transmitted over a range of 10.5 octaves and stored as a sequence of data in the computer. The transmission may be carried out directly from the keyboard, but also be preplayed on the relevant instrument while the computer is on. The computer stores all necessary data in its memory, and this can be accessed or modified as often as needed.

The MIDI system offers a number of facilities via appropriate software.

- The music signal is digitized whereupon note information is passed as a binary code to the computer memory.
- Since the instruments are played live by the computer, the speed and pitch may be varied independently of each other at a later stage.
- Transposing certain fragments of the piece of music enables different keys to be tried out.
- The sound of keyboards may be modified at a later stage.
- Separate patterns may be recombined into a different arrangement.
- Since during playing the instrument are driven directly, that is, without the intervention of a tape recording, there is no loss of quality: the music is played live again and again.

The interface has at least connectors for MIDI IN and MIDI OUT, where in and out mean the direction of the data stream: MIDI OUT sends data and MIDI IN receives data. Most instruments also have a MIDI THRU connection, which enables the MIDI data to be passed to another apparatus.

Connectors on a MIDI unit are invariable 6.3 mm DIN types, so that linking to a mixer is a very simple affair. A typical setup of a MIDI system is shown in Figure 10.4.1. Note that an electric guitar cannot really be controlled via a MIDI, and it is, therefore, played as usual and its signal added to the MIDI signal via the mixer.

The MIDI system has an unfortunate effect on many recordings. It provides a noticeable reduction in labour in the case of keyboard (synthesizer) music and singers who want to be accompanied by such music. This means that there is no longer a need for a band or accompaniment. The commercial music world makes grateful use of this by producing 'music' in the shortest possible time, at the lowest cost, and with the least effort so as to make a maximum profit. The typical disco-type drone and much of the music on recordings and broadcast by commercial radio stations is a sad reflection of this policy.

The MIDI system is not restricted to the sound studio: it offers many facilities for use at home or in the practice area. Once an arrangement has been concluded, all keyboards and a drum computer may be recorded on a single track of a tape recorder. Even if the

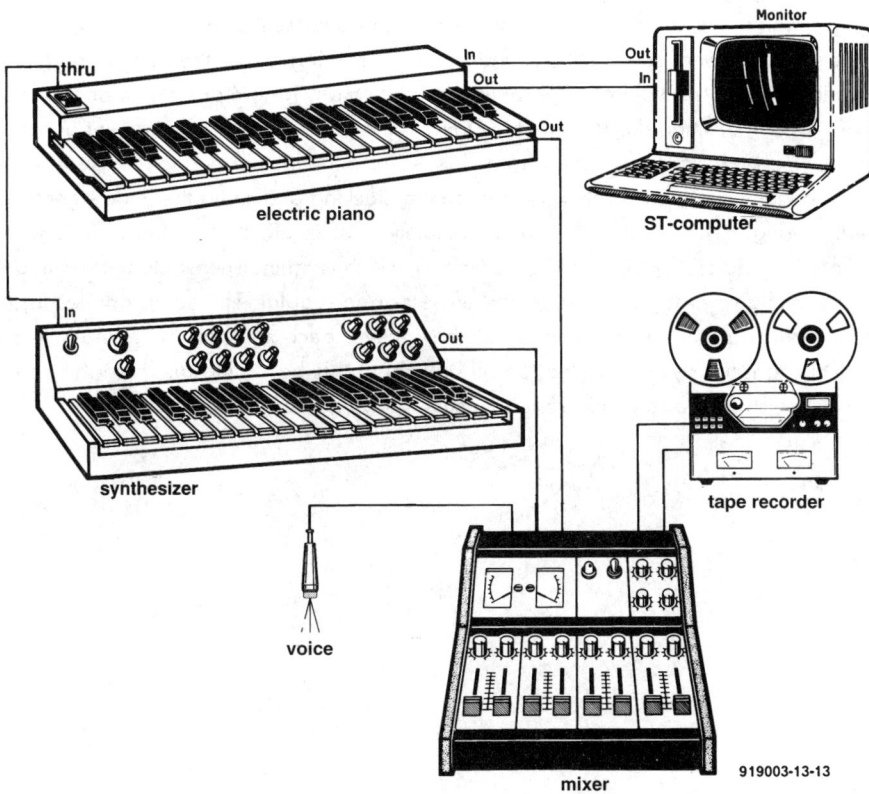

Figure 10.4.1. Setup of a simple MIDI system.

295

available recorder has only four tracks, this still leaves three (monophonic) tracks for, say, vocalist(s) and guitar.

Another blemish of the MIDI system is that the music produced with it is too flat and perfectly timed. Of course, good timing is a prerequisite of good music, but too perfect timing degrades the music to a mathematical series of sounds. This leads to a sterile, dead and often disagreeable, disturbing sound, which, even if only one MIDI-driven instrument is used, is discernible. Why is this?

What we hear is the result of a cooperation between our sensory organ (the ear) and our signal processor (the brain). The perceived sounds are not processed by the brain according to exact mathematical rules: emotion has to play a role. This means that what we hear is not the reality we perceive. One reason for this may well be that we search for what we are used to hear. Natural sounds, such as the chirping of birds, or the murmur of breakers at the seaside, are not simple and regular. For music to sound pleasant and warm, small irregularities are indispensable, but that should not be confused with bad playing. Irregularities in this sense are the extraneous sounds that are produced together with the wanted sound, such as the sound of fingers moving across the strings of a guitar, little variations in pitch, volume, timbre, timing, and so on. Timbre or tone-colour is a real part of music; it distinguishes the quality of tone or volume from one instrument or singer from another, for instance, a flute from a clarinet, or one tenor from another. Also, a guitar solo sounds particularly lively when during the plucking of a string the fingers gently vibrate the string. However, this vibrato immediately loses its effect when the variations are constant and regular. In other words, the vibrato, which is an indispensable part of timbre, must have its own timbre, and must therefore not be regular. The same applies to the wah-wah effect: automatic wah-wah units have never become successful. Finally, many singers subconsciously sing a little too high, without this becoming disagreeable, but it does distinguish the voice better from the accompanying band or orchestra.

Appendix 1

Circuit symbols

—	direct current (d.c.)		light emitting diode (LED)
∿	alternating current (a.c.)		
——	lead or conductor		
	crossing without connection		neon lamp
	crossing with connection (should not be used)		
	removable connection		direct-current motor
	battery		
	capacitor (general) or capacitance		rectifier
	electrolytic capacitor		audio jack (socket) for chassis mounting
	resistor or resistance		

919003-A-1

variable resistor or potentiometer	loudspeaker or drive unit
variable resistor or potentiometer	incandescent bulb
preset potentiometer	press-to-make switch
	press-to-open switch
fuse	
air-cored inductor or inductance	triode (electron tube or thermionic valve)
iron-cored inductor or inductance	
transformer	pentode (electron tube or thermionic valve)
relay	
diode (general) or rectifier	analogue measurement or test instrument
earth or ground	ammeter
microphone	voltmeter

919003-A-2

Appendix 2

Glossary

Absorption
The ability of materials to take energy from the sound field. The degree to which this happens depends on the nature of the material and on the frequency. The converse of reflection.

Accent
In music, a note that is played louder than the others.

Accentuation
Emphasis of a desired band of, usually audio, frequencies.

Access time
The time required by a compact disk player or laser disk player to locate and start playing a selection of a recording after it has been instructed to.

Acoustic
Pertaining to hearing or heard sound produced entirely by physical vibrations without the aid of electronics.

Acoustic coupling
Transmission of information via a sound link, such as between a telephone and a pickup/reproducer in remote computer, control and facsimile operations.

Acoustic feedback
Feedback resulting from sound waves from a loudspeaker or other reproducer reaching the microphone or other input transducer in the same system.

Acoustic filter
Any sound-absorbing or transmitting arrangement which transmits sound waves of the desired frequency while attenuating or eliminating others.

Acoustic frequency response
The sound-frequency range as a function of sound intensity; a means of describing the performance of an acoustic device.

Acoustic impedance
The acoustic equivalent of electrical impedance. The opposition by air to the motion of a surface pushing against it.

Acoustic pressure
The same as sound pressure level.

Acoustics
The physics of sound; the study and application of acoustic phenomena.

Acoustic transducer
(1) Any device, such as headphones or a loudspeaker, for converting electrical audio-frequency signals into sound waves. (2) Any device, such as a microphone, for converting sound waves into alternating, pulsating or fluctuating electrical currents.

Active crossover
An electronic crossover that uses one or more active devices for gain or buffering.

ADC
Analogue-to-digital converter
AES
Audio Engineering Society (of USA).
AF
Audio frequency
All-pass filter
An RC network that introduces a desired phase shift without affecting the frequency response.
Ampère
Unit of electrical current.
Amplitude
(1) Of a sound wave: the maximum deflection of an air molecule. (2) Of an electrical wave: the maximum level of voltage or current.
Analogue quantity
A continuously variable quantity.
Anechoic
Having no echoes or reverberation
ATF
Automatic track find
Audio-frequency signal
A signal with one or more frequencies in the range 20 Hz to 20 kHz
Baffle
Enclosure to prevent leakage of sound waves from the front to the rear of the diaphragm of the drive unit.
Banana plug
Single conductor male connector
Band-pass filter
Circuit that allows a certain frequency band to pass unhindered.
Bass clef
In musical notation, the lower staff of two
Bel
Aid to calculating levels in audio circuits,

named after Alexander Graham Bell, who was the first electroacoustic engineer and inventor of the telephone
Box
Loudspeaker enclosure
CAV
Constant angular velocity
CCW
Counter clockwise; anticlockwise
CD
Compact disk
CD-I
Compact disk interactive
Chorus
Spatial sound effect resulting from an artificially produced beat
Chromatic scale
Musical scale that includes tones and semitones
CLV
Constant linear velocity
Colouration (US: coloration)
General term for an irregular or afflicted frequency response
Compression
Reduction of dynamic range
Concerto
Musical composition for a single instrument and orchestra
Cone
Diaphragm or membrane of loudspeaker drive unit
Counterpoint
The ability, unique to music, to say two things at once comprehensibly; polyphony
Crackle
Random, intermittent clicks
Crescendo
Becoming gradually louder

Crossover network
  Equalizer network that divides the audio
  range into bands
Cut-off frequency
  The frequency where output has declined by
  3 dB referred to the midband frequency
DAC
  Digital-to-analogue converter
Damping
  The antithesis of resonance; gradual
  dissipation of energy
Damping factor
  Figure of merit based on the difference
  between the nominal impedance of a
  loudspeaker and the impedance of the
  source
DAT
  Digital audio tape
dB_{SPL}
  Decibels indicating (acoustic) sound pressure
  level with 0 dB taken as the threshold of
  human hearing
Decade
  Any interval of 10
Decay
  The dying away of a sound from its peak
  value
Decibel (dB)
  One tenth of a bel
Decrescendo
  In music: becoming gradually softer
Deep bass
  Audio frequencies of 20–40 Hz
Diaphragm
  The air-moving surface of a loudspeaker
  drive unit
Digitize
  To transform an analogue quantity into a
  binary number

DIN
  Deutsches Institute für Normung; the
  German standards organization equivalent
  to BSI in the UK and ANSI in the USA
Discordant
  Unpleasant to the ear
Dissonant
  Harsh or unpleasant to the ear
Divider network
  Same as crossover network
DNR
  Dynamic noise reduction
Dome
  Hemispherical diaphragm
Driver, drive unit
  The motor of a loudspeaker
Dynamic range
  The difference between the peak output level
  of a signal and the noise floor
Edit
  To rearrange the order of recorded
  programme segments on a tape
Eigenfrequency
  Frequency of vibration of an audio system
  that vibrates freely
Eigentone
  Natural frequency of vibration of a system
EMI
  Electrical & Musical Industries Ltd
Enclosure
  Housing for a drive unit
Equalizer
  Device to allow the frequency response of an
  audio signal path to be adjusted in some
  way
Expander
  Device to restore the dynamic range of a
  compressed audio signal

Extreme highs
Audio frequencies above 10 kHz

Fader
A control for fading and fading out audio signals

Fidelity
Quality and faithfulness in audio signal reproduction

Filter
A frequency-selective network

Folded horn
A type of loudspeaker whose enclosure has a horn-shaped passageway for enhancing the bass response

Fortissimo
Loud or very loud

Frequency response
A measure of how effectively a circuit or device transmits the different frequencies applied to it

Fundamental frequency
The principal component of an audio wave

Gain
Amplification expressed in dB

Gramophone
Record player

Graphic equalizer
Equalizer that functions simultaneously at a number of preset frequencies

Harmonic
A tone that is an integer multiple of a lower frequency; called overtone by musicians

Headroom
The signal level between zero and clipping levels

Hum
A low-pitched noise in audio systems that consists of harmonically related frequencies

Hump
Rise and fall in a frequency response

Impulse
Brief burst of signal energy

Infinite baffle
An airtight speaker enclosure that totally absorbs or dissipates the rear sound waves of the driver

Infra-bass
Audio frequencies below 30 Hz

Interval
In music, the pitch interval between two notes

Intonation
In music, the tuning or accuracy of pitch

ISVR
Institute of Sound & Vibration Research (UK)

Jack
A socket

Kunstkopf
Dummy head (German) for testing small microphones that take the place of the ears

Leader
Non-recordable beginning and end of a magnetic tape

Linearity
Straightness of transfer function

Loudness
Generally synonymous with volume, which is the intensity of perceived sound

Loudness control
Combined tone and volume control

Low frequency
Audio frequencies below 160 Hz

Low-pass filter
A filter that passes all frequencies below a specified one

Lower highs
The 1300–2600 Hz frequency band

Magnetic tape
  Recording medium used in tape recorders
Master
  The original of a recorded performance
  from which copies can be made
Microphony
  Electrical noise generated by vibration
Midband
  The 500 Hz to 5 kHz band, centring on
  3 kHz where the human ear is most sensitive
MIDI
  Musical Instrument Digital Interface; the
  standard protocol for communication
  between computers and musical instruments
  and vice versa
Mike
  Microphone
Mixer
  Audio control unit for combining two or
  more audio signals into a single, composite
  signal
Monophonic
  Single-channel
Multi-track recording system
  A recording system in which the medium has
  two or more recording paths
Multi-way loudspeaker
  A loudspeaker that has more than one driver
  for the reproduction of different audio bands
Music power
  A disreputable means of specifying the
  power rating of an amplifier
Note
  A single sound of a given pitch and
  duration; called tone in the USA
Octave
  An interval of eight notes, including the top
  and bottom ones

Omnidirectional
  Radiating uniformly in all directions
Oversampling
  A technique where each bit from each
  channel is sampled more than once
Overtone
  The musical equivalent of an harmonic
Pan pot
  A potentiometer used to adjust the stereo
  balance of a monophonic signal allowing it
  to be positioned anywhere across the stereo
  range
Parametric equalizer
  An equalizer whose parameters can be
  varied by the user
Partial
  In music, the same as overtone
Passage
  A section of a musical composition
PCM
  Pulse code modulation
Phon
  The subjective unit for measuring the
  apparent sound pressure level
Phono plug
  Coaxial plug for unbalanced
  interconnections between audio units
Phonograph
  American name for record player
Pianissimo
  Very soft
Piano
  Soft
Pickup
  Any device that converts a sound, scene, or
  other form of intelligence into corresponding
  electric signals
Pink noise
  Noise whose amplitude is inversely

proportional to frequency over a specified
range

Pitch
The property according to which notes
appear to be high or low in relation to each
other

PLASA
Professional Light And Sound Association
(UK)

Polar diagram
A diagram in which the magnitude of a
quantity is shown by polar coordinates

Pole piece
One of the poles of the magnet of a
loudspeaker driver

Power bandwidth
The range of audio frequencies over which
an amplifier can produce half its rated
power without exceeding its rated distortion

Power handling
A measure of the maximum power input a
loudspeaker can absorb without damage or
unacceptable distortion

Q channel
One of two data subcodes recorded at the
beginning of every compact disk

Quantization
Same as digitizing

Quiescent
At rest

Reflection
Audio wave returned by a surface without
change in its wavelength

Refraction
The bending of an audio wave as it passes
from one medium into another

Resampling
Same as oversampling

Reverb(eration)
Tapering prolongation of a sound

RIAA
Recording Industry Association of America

Rolloff
Gradual increase in attenuation over a
range of frequencies; also called slope

Rolloff point
Same as cutoff point

Sampling
Taking samples of an analogue wave at
recurring intervals so that the original wave
can later be reconstructed with reasonable
fidelity from the samples

Sampling rate
The frequency at which samples are taken

SCMS
Serial Copy Management System

Score
In music, a copy combining in ordered form
all the different parts allotted to various
performers

Seamless
Continuous

SMPTE
Society of Motion Picture and Television
Engineers (USA)

Spider
Flexible ring or collar used in a loudspeaker
driver to centre the voice coil on the pole
piece

Subharmonic
A harmonic that is an integer fraction of the
fundamental frequency

Subwoofer
Loudspeaker specifically intended to
reproduce audio frequencies in the
20–100 Hz band

Surround
  The part of the cone of a loudspeaker driver
  by which its outside edge is anchored
Synthesizer
  A device that electronically produces musical
  sounds or acoustic effects
Timbre
  Tone colour
Tone
  Quality of musical sound
Transfer function
  The in/out formula of an amplifier or signal
  path
Treble
  Audio frequencies from 1500 Hz to 20 kHz
Tweeter
  High-frequency drive unit
Velocity of sound
  333 m/s at sea level at 20 °C
Volume unit
  Arbitrary sound level standard related to the
  decibel and used for calibrating recording
  levels
VU
  Volume unit
Weighting
  Correction factor added to a measurement
  to make it correlate more accurately with
  subjective perception
White noise
  Random noise whose amplitude is
  independent of frequency
Woofer
  Loudspeaker intended primarily for the
  reproduction of bass frequencies
XLR connector
  Standard connector for balanced signal lines

Appendix 3

Sound colouring

Aggressive
  Forward and very bright
Airy
  Smooth, effortless high treble
Alive
  Seemingly present in the listening area
Bloom
  Richness and warmth
Body
  Roundness and robustness
Boomy
  Excessive bass emphasis
Boxy
  Hollow
Bright
  Hard, steely
Brittle
  Excess of high, metallic sound
Chalky
  Fine-grained
Clean
  Undistorted
Clinical
  Clean, bright, but mildly pejorative
Closed-in
  Treble lacking above 10 kHz
Coarse
  Very gritty
Cold
  Very cool
Cool
  Light and lean

Crisp
  Peaking at 3–4 kHz
Dark
  Tilting down from the bass upwards
Diffuse
  Confused
Dirty
  Fuzzy or spiky
Dished
  Too much bass and treble
Dry
  Lacking reverberation
Dull
  Lacking in treble
Euphonic
  Pleasing at the expense of accuracy
Fast
  Taut over the whole audio range
Flat
  Departing from correct intonation
Fuzzy
  Having a spiky yet soft texture
Glare
  Distorted mid-treble
Glassy
  Bright
Grainy
  Having excess texture
Gritty
  Like grainy, but harder and coarser
Gutsiness
  Having low, visceral bass

Hard
  Having excessive midrange emphasis
Harsh
  Dissonant, unpleasant, discordant
Heavy
  Having excessive low-frequency energy
Lean
  Having clean, transparent bass
Lifeless
  Superficially perfect, clinical
Loose
  Having badly damped bass response
Lush
  With lots of coherent reverb
Muddy
  Lacking definition
Muffled
  With rapidly reducing high frequencies
Muted
  Lifeless, closed in
Nasal
  With emphasis at 1 kHz
Neutral
  Free from colouration
Open
  Without apparent upper frequency limit
Pinched
  Lacking in bass
Plummy
  Rich, lush
Presence
  centred on 2 kHz
Realism
  Being transparent
Reticent
  Distant-sounding
Rough
  Gritty, harsh

Screechy
  Piercing, unpleasantly so
Sharp
  Strident
Sheen
  With very high treble
Shrill
  Same as screechy
Silky
  Velvety smooth, delicate
Silvery
  Hard, but clean
Slow
  Rhythm seems slow
Smooth
  Easy to listen to
Soft
  Same as closed in
Soggy
  Having ill-defined bass
Sparse
  Lean, but more so
Spiky
  Having very coarse texture
Steely
  Shrill, strident
Sterile
  Same as clinical
Strident
  Same as screechy
Sweet
  Smooth and delicate
Taut
  Under tight control of music signal
Tight
  Well-controlled
Tizzy
  Having excess at 12–16 kHz

Toppish
  Tizzy, but more so
Transparent
  Portraying great realism
Tubby
  Having excessive low bass
Turgid
  Heavy and muddy
Unctuous
  Excessively rich and plummy
Visceral
  Producing a physical sensation
Warm
  Dark, but not so pronounced
Wiry
  Having an edgy treble response
Woolly
  Same as loose

INDEX

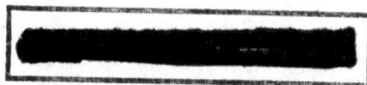